PRAISE FOR
North Pole, South Pole

"A wonderful read that put me in mind of Dava Sobel's books. This is an insightful and lively account of a complex subject that deftly weaves the story of Earth's magnetic field through vignettes of physicists, mathematicians, and explorers through the ages, culminating in the persuasive observations of modern paleomagnetists and theorists."

—DENNIS KENT, Board of Governors Professor
of Earth and Planetary Sciences, Rutgers University

"[An] engaging appreciation of science at work discovering the mysteries of magnetism." —*Kirkus Reviews*

"A fantastic story, highly readable."

—SIMON LAMB, author of *Devil in the Mountain*

"A compelling narrative of the two-thousand-year scientific struggle to unlock the innermost secrets of the cosmic speck of dust we call home. Engagingly written in a lively style accessible to all."

—M. E. (TED) EVANS, Professor Emeritus,
Department of Physics, University of Alberta

"A wonderful, joyful, lucid book. Turner is a natural storyteller."

—TED IRVING, Geological Survey of Canada

THE EXPERIMENT

BECAUSE EVERY BOOK IS A TEST OF NEW IDEAS

"Clearly written and beautifully illustrated."

—SIR PAUL CALLAGHAN, Alan MacDiarmid Professor
of Physical Sciences, Victoria University

"A fascinating read."   —KENNETH CREER, University of Edinburgh

"Gillian Turner has a great gift for writing about science, and personal knowledge of many of the modern giants of geomagnetism. This book will enthuse anyone, young or old, about the physics of the world around them."   —TED LILLEY, Australian National University

"In recent years, many very good books for interested non-scientists have been published: Richard Dawkins's *Climbing Mount Improbable* and *The Ancestor's Tale*, Stephen Jay Gould's *The Lying Stones of Marrakech*, and Dava Sobel's *Longitude* and *The Planets*, to name some of them. *North Pole, South Pole* . . . is a worthy addition to that list. . . . Gillian Turner has a great story to tell, and she tells it well."

—*The Press* (New Zealand)

North Pole, South Pole

North Pole, South Pole

*The Epic Quest to Solve the Great Mystery
of Earth's Magnetism*

Gillian Turner

THE EXPERIMENT
NEW YORK

The Experiment, LLC
260 Fifth Avenue
New York, NY 10001-6425
www.theexperimentpublishing.com

First edition published by Awa Press, Wellington, New Zealand. First North American edition published by arrangement.

The Experiment's books are available at special discounts when purchased in bulk for premiums and sales promotions as well as for fundraising or educational use. For details, contact us at info@theexperimentpublishing.com.

Library of Congress Control Number: 2010934222
ISBN 978-1-61519-031-7

Cover design by Michael Fusco | michaelfuscodesign.com
Typeset by Jenn Hadley, Wellington, NZ
Author photograph by Robert Cross, Victoria University of Wellington Image Services
Cover illustrations: "Magnetic variations at sea," an engraving from *De Magnete* by William Gilbert, 1600 (Science and Society Picture Library). "The Earth's figure and dimensions," a drawing from *The Beauty of the Heavens* by Charles F. Blunt, 1849 (Mary Evans Picture Library).

Manufactured in the United States of America
First printed January 2011
Published simultaneously in Canada

10 9 8 7 6 5 4 3 2 1

To my family—
my parents, my husband, my children

Contents

Illustrations

Main Characters

Ampère, André-Marie (1775–1836). French mathematician and physicist; founder of electrodynamics.

Brunhes, Bernard (1867–1910). French geophysicist who, with Pierre David, discovered lava flows and baked clays magnetized in the opposite direction to Earth's magnetic field.

Bullard, Edward (Teddy) Crisp (1907–1980). English geophysicist; early researcher on dynamo theories of Earth's magnetic field.

Coulomb, Charles-Augustin de (1736–1806). French military engineer who discovered the inverse square laws of electrostatic and magnetostatic attraction and repulsion.

Creer, Kenneth (born 1925). Member of the group of Cambridge paleomagnetists who, in the 1950s, discovered apparent polar wander, contributing to the confirmation of polarity reversals and continental drift; later professor of geophysics at Edinburgh University.

d'Entrecasteaux, Bruni (1739–1793). French explorer who, with Elisabeth de Rossel, made the first measurements of relative geomagnetic intensity.

Elsasser, Walter Maurice (1904–1991). American geophysicist; early researcher of geomagnetic dynamo theories.

Faraday, Michael (1791–1867). English experimental physicist and chemist; director of the Royal Institution; discoverer of electromagnetic induction.

Gauss, Carl Friedrich (1777–1855). German mathematician instrumental in establishing a worldwide network of geomagnetic observatories, and developing first mathematical representation of geomagnetic field.

Gellibrand, Henry (1597–1636). Professor of astronomy at Gresham College, London, who discovered that declination, the angle of deviation of a compass needle from true north, changes with time—the phenomenon known as geomagnetic secular variation.

Gilbert, William (1544–1603). Sixteenth-century experimentalist; physician to Queen Elizabeth I, and author of the classic work *De Magnete*.

Glatzmaier, Gary (born 1949). American solar physicist and geophysicist; with Paul Roberts, developed the first internally consistent computer simulation of a magnetohydrodynamic, self-exciting dynamo in the Earth's outer core to undergo spontaneous polarity reversals.

Graham, George (1675–1751). English compass-maker who discovered the geomagnetic diurnal (daily) variation.

Graham, John American paleomagnetist who, in the mid twentieth century, designed methods to test the authenticity and antiquity of magnetization in rocks; a skeptic of field reversal theory.

Halley, Edmond (1656–1742). English astronomer and explorer who produced the first chart of magnetic declination, covering the Atlantic Ocean, and developed a four-pole theory of Earth's magnetic field.

Hansteen, Christopher (1784–1873). Norwegian astronomer and physicist who advocated and elaborated on Edmund Halley's four-pole theory of Earth's magnetic field, and produced the first charts of geomagnetic intensity (isodynamic charts).

Hartmann, Georg (1489–1564). Vicar of Nuremberg; keeper of early records of declination (circa 1510), and possibly inclination (circa 1544).

Hospers, Jan (1925–2006). Cambridge paleomagnetist who, in the early 1950s, discovered sequences of normally and reversely magnetized lava flows, and proposed the geocentric axial dipole hypothesis. Later professor of applied geophysics at Norwegian Institute of Technology, University of Trondheim.

Irving, Edward (Ted) (born 1927). Member of the group of Cambridge paleomagnetists who, in the 1950s, discovered apparent polar wander, contributing to the confirmation of polarity reversals and continental drift. Established the first paleomagnetism laboratory at the Australian National University, Canberra, and later another at the Pacific Geoscience Center in British Columbia, Canada.

Larmor, Joseph (1857–1942). Anglo-Irish physicist and Fellow of St. John's College, Cambridge University, whose idea of a solar magnetic dynamo eventually led to dynamo theories of Earth's magnetic field; his work on nuclear magnetic moments led to modern nuclear resonance and magnetic resonance imaging techniques, and to a range of magnetometers based on the same principle.

Lehmann, Inge (1888–1993). Danish seismologist who discovered the solid inner part of the Earth's core.

Magnes. Legendary Greek shepherd whose iron-tipped boots and staff were attracted to magnetized lodestone rocks, and whose name is reputedly the origin of the word "magnetism."

Maricourt, Pierre Pèlerin, de. *See* **Peregrinus, Petrus**.

Matthews, Drummond (1931–1997). Cambridge University marine geophysicist who, with his student Fred Vine, first published an explanation of marine magnetic anomalies in terms of geomagnetic polarity reversals and seafloor spreading.

Matuyama, Motonori (1884–1958). Japanese paleomagnetist who first suggested that normally and reversely magnetized rocks follow an age sequence.

Maxwell, James Clerk (1831–1879). Scottish professor of theoretical physics who founded the laws of electromagnetism, invariably known as Maxwell's Equations.

Mercanton, Pierre (1876–1963). French paleomagnetist who showed that reversely magnetized rocks occur all over the globe.

Morley, Lawrence (born 1920). Toronto-based geophysicist whose theory combining seafloor spreading and geomagnetic field reversals to explain observed patterns of marine magnetic anomalies, proposed at the same time as that of Vine and Matthews, was originally rejected for publication; eventually acknowledged as the co-discoverer of the Vine-Matthews-Morley theory.

Neckam, Alexander (1157–1217). English monk from St. Albans; the first European to describe the use of a compass for navigation.

Néel, Louis (1904–2000). French physicist; developed a theory to explain the stable (thermo-) remanent magnetization of volcanic rocks and lavas, carried by tiny grains of magnetic minerals.

Norman, Robert. Sixteenth-century English hydrographer who, in 1576, discovered and described the inclination of the geomagnetic field.

Ørsted, Hans Christian (1777–1851). Danish professor who discovered that an electric current has an associated magnetic field.

Peregrinus, Petrus Thirteenth-century French crusader; author of *Epistola de Magnete* (1269), in which he described the poles of a magnet, the poles of the Earth and magnetic compasses.

Roberts, Paul (born 1929). British theoretical geophysicist who, with Gary Glatzmaier, developed the first internally consistent computer simulation of a magnetohydrodynamic, self-exciting dynamo in the Earth's outer core to undergo spontaneous polarity reversals.

Runcorn, S. Keith (1922–1995). Member of the group of Cambridge paleomagnetists who, in the 1950s, discovered apparent polar wander, contributing to the confirmation of polarity reversals and continental drift; dynamo theorist, later professor of physics at University of Newcastle upon Tyne.

Sabine, Edward (1788–1883). Anglo-Irish scientist and explorer; supervisor of British colonial geomagnetic observatories.

Thales of Miletus (*c.* 624–546 BC). Greek philosopher and reputed founder of many branches of mathematics and science, including electricity and magnetism.

Thellier, Émile (1904–1987). French experimental paleomagnetist who, with his wife Odette Thellier, worked extensively on the determination of the intensity of the paleomagnetic field from lavas and archaeological artifacts.

Vine, Fred (born 1939). Cambridge geophysicist who, with Drummond Matthews, was the first to publish a theory combining seafloor spreading and geomagnetic field reversals to explain the patterns of marine magnetic anomalies.

Von Humboldt, Alexander (1769–1859). Prussian naturalist and explorer who was the first to recognize the magnetization of rocks, and to publish measurements of the (relative) intensity of the geomagnetic field.

Introduction

If there is anything I share with the twentieth-century genius Albert Einstein, it is a fascination with the magnetic compass. It was this that led both of us to one of the greatest problems in physics: finding the origin of the force that draws the compass needle unerringly towards the north. The story I am about to tell has grown from a notion sown in my mind by my publisher in 2005, the International Year of Physics and the centenary of Einstein's *annus mirabilis*.

Einstein became entranced by magnetism at the age of seven, when his father gave him a compass. For me the moment of truth struck during a short series of lectures in my third year of undergraduate physics at Cambridge University. Cambridge had a curious assessment exercise, the "prepared essay" exam. By the time my essay, "Our Magnetic Planet," was honed and practiced, no one could have crammed more information into the two hours of solid writing we were allowed. I was well and truly hooked.

A long journey of discovery lay ahead. It took me first to Edinburgh, where as a research student I mucked about in small boats collecting long cores of lake-bottom mud from such beauty spots as Scotland's Loch Lomond and the Lake District's Windermere. From my lake mud I uncovered the wanderings and variations of Earth's magnetic field that had lain hidden for 10,000 years. I shared an office with Stavros, a Greek ex-army officer; Eric, a sports fanatic who somehow fitted in his research between seemingly end-less games of squash, tennis, soccer and badminton; Julie, a quiet, industrious student, who zipped around town on a tiny scooter; and Ruth, who, were it not for her meager budget, would have owned a full-blown Harley-Davidson. Together we ruled the roost in the James Clerk Maxwell Building, taunted the professor, blissfully unaware of his pre-eminence in the foundations of our subject, and gradually unraveled the secrets of geomagnetic secular variation—the curious way in which Earth's magnetic field keeps changing.

In those early days I found the notion of polarity reversals—that the magnetic field could, and had, turned right upside down many times—quite incredible. Like a doubting Thomas, I had to put my fingers into the holes left in the rock once my samples had been drilled out, then go through the necessary magnetic measure-ments to be convinced.

There was now no turning back. I went to conferences and met the pioneers of the subject: Runcorn, Creer, Lowes, Tarling and Gubbins. I moved to a post-doctoral fellowship in Canada and met more: Irving, Cox, Dalrymple, McDougall, and eventually Roberts and Glatzmaier. I am proud to count them among my colleagues and friends, and it is my privilege to take you on a journey to meet them, and many of the earlier explorers of geomagnetism.

Have you ever wondered at the sheer uniqueness of Earth, the amazing coincidences of physics and chemistry that enabled life to flicker into existence here, to take hold and flourish on a tiny speck of dust, bound in orbit around a little ball of fire, floating through the vastness of the universe?

Earth is just the right distance from the sun to carry water as solid, liquid and vapor: glaciers, oceans and clouds. If it were closer to the sun and hotter, ice caps and glaciers would be unknown. If it were much further from the sun, it would be frigid and inhospitable. Our oxygen-rich, life-sustaining atmosphere is retained by a delicate balance of gravity and temperature: a moderate greenhouse effect ensures that most of us enjoy a cozy environment for most of the year, while a tiny amount of ozone high in the stratosphere protects us from hazardous ultraviolet rays beating in from the sun.

The atmosphere is, however, perilously thin. You could compare it to the thickness of paper covering a school globe. And interplanetary space is anything but empty—it streams with charged particles and radiation that are harmful to human health. The sun throws out protons and electrons in all directions at speeds of hundreds of kilometers per second, while even more energetic particles bombard the solar system from outer space. This solar wind and these cosmic rays would make Earth quite uninhabitable were it not for the fact that we sit in the middle of an enormous magnetic shield that arrests and diverts the onslaught way above our heads.

Most of the solar wind and cosmic ray particles are safely deflected by Earth's magnetosphere and continue uninterrupted on their race into outer space. Only the most energetic penetrate the shield and these become trapped, forced into spiral paths around Earth's magnetic field lines, bouncing back and forth from pole to pole in the so-called van Allen radiation belts. Occasionally an especially energetic burst gives rise, at high latitudes, to a shimmering show of lights—the *aurora borealis* or *aurora australis*.

The sun itself is strongly magnetic, and so are the four giant outer planets, Jupiter, Saturn, Neptune and Uranus, but amazingly Earth is the only one of the inner planets to have a strong magnetic field. Why should it be different from its neighbors, Venus, Mars and Mercury? Along with the rest of the solar system, all four are thought to have formed around the same time, 4.5 billion years ago, and all have similar interiors: a dense metallic core, a less dense mantle and a rigid, rocky crust. So why is only Earth magnetic?

For millennia, magnetism has commanded a magical sort of curiosity. The Greeks were mystified by the attractive properties of lodestone, magnetized rock. The first Chinese compasses were astrological instruments, used to divine the ways of the winds and waters. Even after the mariner's compass had become an essential tool of navigation, magnetism retained its fascination and became a challenge to explorers and scientists alike. No wonder medieval scholars placed the source of the compass needle's "virtue" in the heavens.

The history of geomagnetism is strewn with famous people and their colorful stories. It was the ever-practical William Gilbert who reasoned that Earth itself was magnetic. Between spotting comets and charting the stars, Edmond Halley was the first to plot Earth's magnetic field accurately, while Karl Friedrich Gauss lent his mathematical genius to analyzing and understanding it. The crucial connection between electricity and magnetism was first discovered and then explored by several great nineteenth-century physicists, including Hans Ørsted, André Marie Ampère, Michael Faraday and James Clerk Maxwell. Together with a greater understanding of the Earth's interior that came with the birth of geology and then geophysics, this eventually focused interest on planetary dynamos. Meanwhile, the geomagnetic field became much more amazing with the realization that it had flipped polarity, and not just once but many many times throughout geological history.

By 1950 the quest to understand Earth's magnetism was focused on a hydromagnetic dynamo, but the mathematical calculations were intractable. Not until the advent of the supercomputer could the equations be finally solved, and scientists present the world with evidence of the gigantic powerhouse heaving in the core of our planet.

However, to begin at the beginning we must travel back two and a half millennia, to ancient Greece.

The Mystery of Magnetism

*Just as there are in the heavens two points more note-
worthy than all the others . . . so also in this stone . . .
there are two points, one north and the other south.*

—PETRUS PEREGRINUS, 1269

Old Magnes had come this way so many times before that his feet
knew every one of the black rocks over which he was clambering.
Day after day, year after year, for most of his life he had trudged
up this hillside to tend his small flock of sheep. Never before,
though, had his iron-studded boots stuck to the rocks the way they
were doing now. The only way he could get his foot free was with
a mighty kick—and then, with his next step, his boot would be
sucked down again.

His staff also seemed to have taken on a life of its own. Each
time he planted it on a rock to steady himself he had to tug hard to lift
it again. What was going on? Last night there had been terrifing

thunder and lightning, followed by torrential rain—thank the gods he had come down to his shelter and not spent the night on the mountain—but the ground was now dry again and everything looked normal.

This legend of the Greek shepherd Magnes is thought to date back to around 900 BC, but it was recorded almost a millennium later by the Roman scholar and writer Pliny the Elder. Pliny was fascinated by the world around him, and before being killed by poisonous gases from Mount Vesuvius during the AD 79 eruption that destroyed Pompeii, he spent much of his life recording his observations of nature in a multi-volume encyclopedia, *Naturalis Historia*. Pliny's story of Magnes, although no doubt embellished through centuries of retelling, provides two important clues to understanding Earth's magnetic properties: an electrical storm took place and rocks became magnetized.

Magnes was apparently climbing on Mount Ida—the same Mount Ida from which Zeus is said to have watched the sacking of Troy—in the northwest of what is now Turkey. This is not far from the region the ancient Greeks called Magnesia after their homeland in mainland Greece. Still today, Magnesia is well known for its deposits of lodestone, a rock that is rich in magnetite, an oxide of iron. Normally a lump of magnetite-bearing rock is unremarkable. However, if the rock is struck by lightning it becomes strongly magnetized. A bolt of lightning may pass an electric current of up to a million amps into the ground—not for long, but long enough for rocks within a short distance to become magnetized intensely and stably.

For centuries magnetism was thought to be unique to lodestone. What was it about lodestone, and only lodestone, the ancients wondered, that gave it this magical property? The earliest ideas on the nature and origin of magnetism are usually attributed to a Greek philosopher, Thales (*c.* 624–546 BC), who lived in Miletus, a busy trading city not far from Mount Ida. Together with his well-known

Thales of Miletus. A Greek philosopher who lived around the sixth century BC, Thales was puzzled by the way lodestones could attract each other and pieces of iron across empty space, and decided that, like humans, they must have souls.

contemporary Pythagoras, Thales is credited with having laid the foundations of not just philosophy but also physics and mathematics. None of his original writings seem to have survived, but Aristotle reported:

> Thales . . . held soul to be a motive force . . . he said that the magnet has a soul because it moves the iron.

The Greeks recognized that lodestones did not attract only other lodestones: they also attracted pieces of metallic iron. And they had observed that a piece of iron in contact with a magnet became magnetized itself, and so was able to attract another piece of iron—a process now known as induction.

Further, a lodestone did not need to be in physical contact with another lodestone or a piece of iron in order to attract it. This "action-at-a-distance" effect, where a force acted across empty space in which no intermediary medium existed, seemed impossible to explain in material terms, so Thales reasoned that an animistic explanation was called for. Living bodies moved, and instilled motion in other material objects. Living bodies had souls. Therefore, in order to move a piece of iron the magnet, too, must possess a soul.

Thales was also familiar with another action-at-a-distance effect, namely that when a piece of amber was rubbed with fur it could attract scraps of chaff and other light particles. (This is the same "electrostatic" effect that makes our hair crackle and stand on end after brushing it on a dry day.) However, whereas rubbed amber attracted scraps of all kinds of materials, lodestone attracted only other lodestones or iron.

These action-at-a-distance effects—which, as well as magnetic and electrostatic forces, also include gravity—would challenge not just Thales. Down the ages, scientists, philosophers, teachers and students would struggle to understand them, and create many and varied explanations.

Later Greek philosophers opted for an "atomistic" view of matter. This bore little resemblance to modern atomic theory, other than the idea that matter was made up of innumerable tiny particles. In the fifth century BC, Diogenes of Apollonia maintained that a lodestone or magnet "fed" on atoms of iron. Another school of thought believed that a magnet emitted particles, and that these particles cleared the space between it and a piece of iron, thus drawing them together.

This last idea led, over ensuing centuries, to a whole host of "effluvia" theories involving invisible emissions from magnetic materials, and finally, in the nineteenth century, to the notion of the magnetic field. At this early stage, though, few theories addressed, let alone answered, the obvious question—why was magnetism confined to lodestone and iron?

Early Greek science was essentially limited to the observation of natural phenomena and endless philosophizing as to their causes. Without the modern elements of prediction, experimentation and testing, alternative theories such as animist, atomist and effluvia could not be evaluated against each other in any substantial way, and so little progress was made.

At the same time as the Greeks were holding sway in the Mediterranean, an advanced civilization was thriving in China. While science there was also inextricably mixed up with mysticism, divination and religion, technology reached a degree of sophistication that would be unparalleled in the West until the Renaissance of the fifteenth and sixteenth centuries.

The earliest recorded compass, a Chinese divining instrument, probably dated from the first century AD, although it could have been in existence as much as 300 years earlier. From early Chinese writings studied and described by the English historian of science Joseph Needham, we know that this compass was used to determine the directions favored by the winds and waters, and so was a guide to laying out villages, building houses, plowing fields, orienting tombs and much more—the ancient art of *feng shui*.

The instrument consisted of a spoon-shaped piece of lodestone, known in China as *tzhu-shih* or "loving stone," which represented the star constellation of *Ursa Major*, the Great Bear. This was delicately balanced on a circular "heaven" plate made of bronze or wood, which was itself placed on top of a square "Earth" plate. Both the heaven plate and the Earth plate were intricately engraved with astronomical symbols and directions. The "spoon" took on a

A reconstruction of the earliest Chinese magnetic compass, which was used to determine the favored directions of *feng shui*. A south-pointing lodestone spoon is balanced on a bronze Earth-plate, which is engraved with Chinese astrological symbols. Kaifeng, Henan Province, China.

natural magnetization along its length so that, when balanced, its handle came to rest pointing to the south. Interestingly, the early Chinese routinely chose south as the prime cardinal direction.

The Chinese seem never to have questioned the nature of the force that aligned their compass. To them, as to the Greeks, such things lay in the lap of the gods. There is, however, documentary evidence that they recognized discrepancies between the compass's south and true south. Between about AD 720 and 1086, Chinese compasses appear to have deviated by up to 15° east of true south, while all later records show the deviation to have been to the west of true south. Indirect evidence of this deviation is to be found in the streets of many ancient Chinese towns and cities, including Beijing and Nanking. A plan of the southern part of the township of Shandan in Gansu province on the Old Silk Road shows two distinct street orientations. The older is due north–south, but the younger deviates by eleven degrees, trending from 11° west of south to 11° east of north. Presumably the streets were aligned to the favorable directions of the winds and waters as determined by the spoon-shaped compass, and between the two periods of building the compass had shifted to the west by eleven degrees.

These early Chinese were not great seafarers or travelers—had they been, the compass would almost certainly have become a navigational tool much earlier. As it is, the earliest reference to a mariner's compass comes from the beginning of the twelfth century. By then the Chinese had perfected techniques for magnetizing a fine iron needle by stroking it with a piece of lodestone and balancing it on a finely made pivot, floating it on water, or suspending it from a fine silk thread in order to minimize the effect of friction and improve its overall performance. Beautiful floating fish and turtle-shaped pivoted compasses originate from this period.

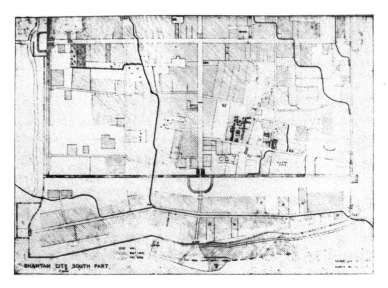

A plan of part of the Chinese town of Shandan, showing streets aligned in two different orientations, one due north–south, the other trending from 11 degrees west of south to 11 degrees east of north. The difference is thought to have come about because of a change in orientation of the magnetic compass between the two periods in which the town was built.

Following the so-called Dark Ages, the compass eventually surfaced in Europe in the writings of an Englishman, Alexander Neckam. Born in St. Albans in Hertfordshire in 1157 on the same night as Richard I, Neckam had grown up with the future king as a foster brother. He went on to teach arts at the newly emerging University of Paris and later returned to St. Albans School before becoming a canon and abbot of the Augustinian abbey at Cirencester. Neckam's interests were far-ranging, from theology to natural philosophy, but he is remembered mainly for his two books, *De Nominibus Utensilium* (*On Instruments*), published around 1180, and *De Naturis Rerum* (*On the Natures of Things*), around 1200.

Each contained an article on nautical navigation. In the first, Neckam explains the use of a magnetic compass needle for navigation at sea, while in the second he extols the advantages of a pivoted needle:

> The sailors moreover, as they sail over the sea, when in cloudy weather they can no longer profit by the light of the sun, or when the world is wrapped up in the darkness of the shades of night, and they are ignorant as to what point of the compass their ship's course is directed, they touch the magnet with a needle. This then whirls round in a circle until, when its motion ceases, its point looks direct to the north.

How the compass had reached Europe is something of a mys-tery. Neckam probably came across it first in Paris, but the tenor of his writing suggests that by the end of the twelfth century it was already in common use by mariners. This is at odds with the suggestion that it was Marco Polo who brought it back to Europe from China. Polo did not visit China until 1275 and returned to Venice in 1295, a whole century after Neckam's descriptions.

Another more persuasive theory is that the compass arrived in Europe courtesy of Arab traders. The presence of ancient Chinese objects through the Persian Gulf and Red Sea regions and along the east coast of Africa indicates that, from the eighth century onwards, there was busy trading between the Arab inhabitants and the Chinese. It is hard to imagine that the compass would not have eventually become an item of trade, and found its way north and west to Europe. The timing—one hundred years from the compass's invention in China to its appearance in Europe—is plausible. However, to complicate the theory, the earliest Arabic references to the compass also seem to post-date Neckam's. It is not until the mid thirteenth century that Arab documents and stone tablets mention sailors finding their way by means of floating compasses fashioned from fish-shaped pieces of iron rubbed with a magnet.

Perhaps the likeliest explanation is that the European compass was developed independently. This is supported by the difference in prime direction: to this day Chinese compasses are made with the prime end of the needle pointing south, while European compasses have always pointed to the north.

Neckam's reports and dates are backed up by a satirical poem, "Bible," written by Guyot de Provins, a French poet and monk, around 1205. It includes the passage:

> . . . there is an art which the sailors have, which cannot deceive.
> They take an ugly brown stone, the magnet, to which iron
> willingly attaches itself, and touching a needle with it, they fix the
> needle in a straw, and float it on the surface of water, whereupon
> it turns infallibly to the Pole Star.

Since Thales' time it had been known that a magnet attracted a piece of iron, while two magnets would attract or repel one another.

A compass needle was just a magnet so what attracted or repelled it, and what caused it to rotate into a north–south alignment?

An interesting explanation emerged. Until the invention of the compass, the heavens had provided the chief means of navigation— the sun by day and the stars by night. It was commonly believed that the Earth lay at the center of Creation, with the moon, sun, stars and known planets—Mercury, Venus, Mars, Jupiter and Saturn— arranged on crystal spheres of increasing size, each of which revolved around the Earth daily. Beyond Creation lived God in his heaven, and beyond this lay infinite space. One star, however, seemed to remain fixed in place, because it lay on the axis about which the celestial sphere of the stars revolved. And, as de Provins pointed out, it was towards this star that the compass needle infallibly turned. Hence, the directivity of the compass came to be attributed to the Pole Star.

An elaboration of the Pole Star theory appears in a poem by a thirteenth-century Italian, Guido Guinicelli:

In what strange regions 'neath the polar star
May the great hills of massy lodestone rise,
Virtue imparting to the ambient air
To draw the stubborn iron; while afar
From that same stone, the hidden virtue flies
To turn quivering needle to the Bear
In splendour blazing in the northern skies.

This brief verse captured several important ideas. Although all eventually turned out to be wrong, they marked significant steps in scientific reasoning. The first was the notion that there were lodestone mountains at the Earth's poles. Since lodestone was the only material Guinicelli knew that attracted a compass needle, he imagined there must be an enormous mass of it at the spot on Earth towards which all compasses were known to point—directly

The medieval concept of a geocentric universe, as depicted by Peter Apian in *Cosmographia*, 1524. The moon, inner planets, sun and outer planets were all thought to revolve around the Earth. Beyond these bodies were the stars, the heavens and God himself.

beneath the Pole Star. The notion of gigantic magnetic mountains at the poles spawned fantastic legends: apparently the mountains could even pull iron nails from passing ships.

Also captured in Guinicelli's verse was the idea that magnetic attraction, or "virtue," was somehow transported through the air between the lodestone and the compass needle. Was this merely fanciful poetic language or an early glimmer of the concept of magnetic fields?

Just a few years later the whole way in which men studied nature was to take a new turn, thanks to a little-known Frenchman and his investigation of magnets. Pierre Pèlerin de Maricourt is believed to have been a knight and a crusader. Commonly known as Petrus Peregrinus, or sometimes as Peter the Wanderer, he was also a military engineer, well educated and something of a scholar. In 1269, while serving in the army of Charles d' Anjou at the siege of Lucera in southern Italy, Peregrinus had found time to reflect on and write about experiments he had earlier carried out. The result was *Epistola de Magnete* (*Letter on the Magnet*), dated August 8, 1269. Addressed to Sygerus de Foucaucourt, Peregrinus's neighbor in Picardy, *Epistola de Magnete* has been lauded as Europe's first work of true science.

Peregrinus had introduced the one crucial element missing from previous scientific endeavors: the idea of learning from experimentation. A contemporary, Friar Roger Bacon, himself a progressive and vigorous proponent of experimental science—he devised methods of making gunpowder, spectacles, mechanical flying-machines, ships, carriages and much more—would describe him as a master of experiment and one of only two perfect mathematicians, the other being one of Bacon's own students.

Peregrinus's most important discovery was magnetic polarity— the distinction between the north and south poles of a magnet. In his experiments he had explored the magnetic properties of a sphere of lodestone, and what happened when a needle or a short iron wire was placed at various points on its surface. He had found two unique locations at which a needle would align itself perpendicularly to the surface of the sphere. These were at opposite ends of a diameter, and the needle was attracted more strongly to these locations than to any others. Further, when the needle was positioned at other places on the sphere its orientations outlined a series of meridians—"just as all the meridian circles on the globe meet in the two opposite poles of the world," Peregrinus wrote. For him, Earth's poles represented the axis about which all the celestial spheres—indeed the whole of Creation—revolved. The "world" extended beyond the globe out into space: Earth was merely at its center.

Peregrinus's recognition of the similarity between the magnetism of the lodestone sphere and Earth's magnetism led to the name "terrella," or "little Earth," being used to describe a lodestone sphere, and north (seeking) and south (seeking) labels being given to the poles of a magnet.

Peregrinus was not finished yet. He now showed that whereas magnetic attraction occurred between the north pole of one magnet and the south pole of another, two north poles or two south poles repelled each other. This fundamental law of magnetism, which had eluded the ancient philosophers, had at last been discovered and enunciated: like poles repel, unlike poles attract.

He then went on to debate why the compass needle should point to the north. First, he argued against the overriding influence of polar lodestone mountains. Even if such mountains existed—and in 1269, since no one had ventured anywhere near the north or south poles, there was no direct evidence either for or against—there were, he pointed out, known lodestone deposits in other places around the globe and these did not influence the compass needle, other than very locally.

His observations of the symmetrical pattern of the needle's orientations around his terrella led him to reason that the principal influence acting on the compass needle had to have the same symmetry, be of global proportions, and dominate any irregular influences due to local lodestone deposits or the possible effects of hypothetical lodestone mountains. In summary he wrote:

> . . . it is from the poles of the world [that is, the universe] that the poles of the magnet receive their virtue.

The Petrus Peregrinus Medal of the European Geosciences Union, depicting Peregrinus's sketch of a "continually moving wheel of wonderful ingenuity," never realized in practice. An inventor of magnetic instruments, this thirteenth-century French knight and scholar is best known for discovering the difference between the north and south poles of a magnet, and that like poles repel one another, while unlike poles attract.

Peregrinus had also noticed that the Pole Star, which could be seen from the northern hemisphere, was not absolutely fixed at the celestial pole. Instead, it moved very slightly around it—a movement the compass needle failed to follow. He reasoned, therefore, that the magnetic "virtue" of the compass could not be attributed directly to the Pole Star either.

The final pages of *Epistola de Magnete* contain Peregrinus's designs for three magnetic instruments. The first two are floating and pivoted compasses; in 1868 an Italian engineer named Bertelli would construct working instruments using these plans. However, neither he nor Peregrinus were able to make the third instrument work. Like all subsequent attempts to build a perpetual motion machine, Peregrinus's "continually moving wheel of wonderful ingenuity" failed hopelessly to live up to his expectations. This was, he claimed, "through lack of skill, rather than a defect of nature." In 2005 the European Geosciences Union honored Peregrinus by creating the Petrus Peregrinus Medal, to be awarded annually for outstanding scientific contributions in magnetism and geomagnetism. In a display of eternal optimism, the medal depicts Peregrinus's "continually moving wheel of wonderful ingenuity."

Peregrinus had certainly come a long way with his concept of Earth's magnetism as perfectly symmetrical about an axis through the poles. That he meant the celestial poles rather than the poles of Earth does not matter—the effect was the same: wherever a compass needle was put on the surface of his terrella, or little Earth, it would point along the meridian or great circle from south to north. What he did not know was that the Chinese had already recorded discrepancies from this perfect picture.

Voyages of Discovery

Who would of his course be sure, when the clouds the
*   sky obscure,*
He an iron needle must in the cork wood firmly thrust.
Lest the iron virtue lack rub it with the lodestone black,
In a cup with flowing brim let the cork on water swim.
When at length the tremor ends, note the way the needle
*   tends;*
Though its place no eye can see—there the polar star
*   will be.*

—WILLIAM THE CLERK, *c.* 1230

According to the Renaissance philosopher Francis Bacon, three things, all of them first invented in China, drew the Western world out of the cultural doldrums of the Dark Ages: the compass, gunpowder and the printing press. With the first of these, the compass, came unprecedented opportunities to explore and chart the world's oceans, and to discover and eventually colonize new lands. Before its advent, mariners had often been forced to navigate along known coastlines.

However, even with a compass to set a course, it was still difficult to determine a ship's absolute position in the open ocean. You could find your latitude by measuring the altitude of the midday sun or the Pole Star with an astrolabe, an instrument dating back over two thousand years to the ancient Greek astronomers. However, longitude could be estimated only by dead reckoning—that is, by assuming your speed and direction from a previously known position. To use the stars to determine longitude you needed to know the time at the home port or some other known location, and sufficiently accurate timekeeping was not available. Even well into the seventeenth century it was difficult to avoid significant errors; for example, the positions logged by Abel Tasman when he sailed up the west coast of the North Island of New Zealand in 1642 erred by some two to four degrees of longitude, and put his ship in the center of the island.

Within a hundred years of Neckam's first reference to the compass, a series of nautical charts had begun to appear, based on the magnetic compass bearings between different ports. These portolan, or harbor-finding, charts showed only coastlines and coastal ports: details of the hinterland areas were not necessary for navigation at sea. From each port a series of straight "rhumb-lines" radiated out in the magnetic directions of other ports. A mariner setting a course from one port to another simply needed to set his chart to his compass and sail in the direction of the appropriate rhumb-line.

Of the 130 or so portolan charts that survive today, most were produced in Italy, Catalonia and Portugal between the fourteenth and sixteenth centuries, and covered the Mediterranean Sea and the Atlantic region to Ireland, and the west coast of Africa. They were originally drawn by hand on sheepskin or vellum, but with the advent of line-engraving on paper in the sixteenth century they could be more easily reproduced. Beautifully engraved and colored compass roses became a characteristic feature.

The shapes of coastlines on portolan charts often appear distorted when compared with modern maps. This is partly because

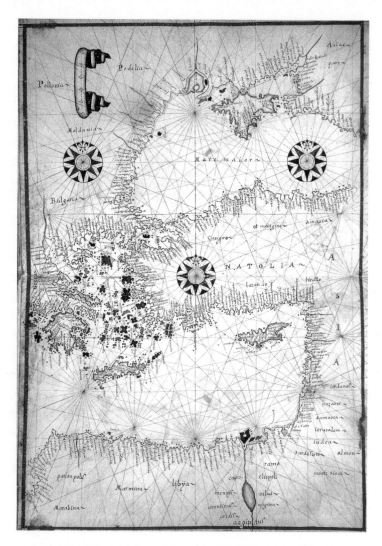

A medieval "portolan" chart of the eastern Mediterranean and Black Sea region, showing compass bearings, or rhumb-lines, between numerous harbors. Sailors simply needed to set their charts to the compass and sail in the appropriate direction. This chart has been attributed to Joan Rizo Oliva, a Catalan chartmaker who lived from 1580 to 1615.

modern map projections compensate for latitude, but also because of the deviation of compass needles from true north. This phenomenon, which came to be called "declination," is the angular difference between magnetic north and true north—the same phenomenon unwittingly recorded by the Chinese in their street alignments in Shandan. Although declination was not widely recognized or understood by the makers of portolan charts, it did not affect the charts' accuracy for navigation since both the chart-makers and the sailors worked from the same compass bearings.

The realization that more often than not compass needles deviated from true north came gradually in Europe. Peregrinus's contemporary Roger Bacon apparently referred to the fact that his compasses rarely pointed exactly along the meridian, and from the beginning of the fourteenth century this came to be noticed more and more frequently. Navigators and philosophers began to query whether the direction-seeking properties of lodestones might depend on their source and, if so, whether such differences, or some other systematic error, could be transferred to needles during the magnetization process. Or might the discrepancy simply be due to careless observation?

Only when these possibilities were satisfactorily refuted was declination accepted as a real feature of Earth's magnetism—a feature for which allowance had to be made whenever a compass was used for accurate direction-finding. German sundials made in the middle of the fifteenth century often incorporated compasses to help with alignment, and some of these had a mark that accurately indicated the adjustment necessary to allow for declination. Similarly, some road-maps produced in Germany showed an image of a compass in the margin, indicating a declination of 11°15′ E and instructions for setting the map accordingly, exactly as hikers, back packers and boy scouts do today.

The discovery of declination meant that Peregrinus's perfectly symmetrical picture of Earth's magnetism was no longer strictly accurate. In the mid sixteenth century, when declination measurements were still relatively few and came mainly from Europe and the Atlantic region, it seemed the discrepancy was small and might be explained by a slight modification of Peregrinus's model.

Gerardus Mercator, a Flemish cartographer, had realized that differences in declination from place to place were also responsible for the curious distortion of coastlines on portolan charts. He figured the effect could be explained by supposing that the magnetic axis of the Earth was tilted with respect to the rotation axis—in other words, if the north and south magnetic poles were offset from the north and south geographic poles. If this were the case, the observed compass directions would still describe a series of "magnetic meridians," but they would intersect at the magnetic poles, not the geographic poles. He estimated the north magnetic pole to lie at 85° N, at a longitude roughly 180 degrees from the Azore Islands (or 151 degrees east of Greenwich). At this time a compass in the Azore Islands pointed due north, meaning that the declination there was zero. Mercator and certain other mapmakers of the time chose this as the prime meridian from which to measure longitude.

Mercator was soon to discover, however, that his "magnetic meridians" did not intersect at single points at all—the declination was too variable and too complicated for that. In a map of the Arctic region that he drew about 1569, but which was not published until after his death in 1594, Mercator opted for not one but two magnetic north poles, one at the geographic pole and one at approximately 150° E and 75° N. Interestingly, he depicted both poles as magnetic islands. Was Mercator an adherent to the lodestone mountain theory or was he guessing? His mapping of the Arctic region was, after all, pure speculation since next to no polar

Map of the *Septentrionalium Terrarum* (Northern Lands or Lands of the
Seven Stars) by Gerardus Mercator, published posthumously in 1595.
It shows a *Rupes Nigra*, a black precipice of lodestone, at the north pole
and a second magnetic island in the yet-to-be-discovered Bering Strait,
as well as many other mythical features.

exploration had actually taken place. California, which had recently
been claimed by Spain, appeared north of the Arctic Circle.

By 1500 the Spanish and Portuguese voyages of discovery had
begun in earnest, with ships sailing east around the Cape of Good
Hope and west across the Atlantic Ocean, often making land
more through luck than navigational skill. As John Cabot, his son
Sebastian Cabot, Christopher Columbus, Vasco da Gama and
Ferdinand Magellan sailed the high seas, the question of determining

longitude at sea became a serious problem, and one that would persist for several centuries.

Over this time many possible solutions were put forward, some ingenious, some almost practical, others quite incredible. Proposed solutions generally fell into one of two categories: the first were based on astronomical observation, the second mostly on magnetic declination. Astronomical observation was eventually to prove the more fruitful.

In principle, longitude, like latitude, could be determined from the stars: you simply needed star charts that referred to a known place—say, your home port—and a clock set accurately to the time at that place. The difference in the positions of the stars between your location and your home port at a particular time would tell you how far around the globe you had traveled and, therefore, the difference in longitude between your location and your home port.

However, although in theory the method was simple, in practice no one had come up with a clock robust enough to keep time to a sufficient degree of accuracy over long sea voyages. (It would not be until 1773 that a Yorkshire clockmaker, John Harrison, having constructed a rugged and reliable chronometer, would claim the £20,000 prize that had been offered in desperation by the British Board of Longitude some sixty years earlier.)

One of the earliest ideas about how to determine longitude based on the compass was conceived by a Neapolitan scholar, Giovanni Battista Della (John Baptista) Porta, and described in his encyclopedia *Magia Naturalis* (*Natural Magick*), published first in 1558 and later, in twenty volumes, in 1589. Porta appears to have examined some early observations of declination. In 1492, for example, Christopher Columbus's navigators had been alarmed to find that on occasion their compasses deviated significantly from the Pole Star. Columbus had apparently explained the effect to his crew in terms of some previously undetected motion of the Pole Star, but Porta supposed that the compass needle had moved in step with Columbus's progress across the Atlantic Ocean, from some ten degrees east of north in Europe to a similar angle west of north by the time he reached the Caribbean region.

Porta concluded that the declination of the compass needle provided a direct measure of longitude east or west of the Azores, where it pointed true north. He even suggested making a giant compass, about five feet in diameter, and dividing the degrees and minutes into seconds and thirds in order to resolve longitude more precisely. Porta's idea was short-lived, but the notion of a meridian of zero declination through the Azores survived a little longer, thanks to Mercator and other sixteenth-century cartographers who still ran the prime meridian through these islands.

Later, and more significant in the quest to understand Earth's magnetism, was the discovery of what is now called "inclination." This is a simple concept. Anyone who has traveled widely with a compass will have discovered that a compass that is balanced at one location tilts up or down when used at other latitudes. This is particularly noticeable between the hemispheres. If a compass designed for use in the northern hemisphere is taken to the southern hemisphere, the north end of the needle seems to be pulled upwards and the south end pushed down. The opposite is true of a southern hemisphere compass used north of the equator: the north end of the needle is pulled down and the south end pushed up. For this reason, a compass needle is usually weighted or balanced to make it swing in a horizontal plane.

If a magnetized but unweighted needle is freely suspended about its center, the angle it takes above or below the horizontal was known historically as the "dip." Today it is called as the "inclination." By convention, inclination is deemed positive if the direction is *below* the horizontal (as at most locations in the northern hemisphere), and negative if *above* the horizontal (as at most locations in the southern hemisphere).

English hydographer Robert Norman's demonstration of the inclination, or dip, of a magnetized needle that is balanced and neutrally buoyant in a goblet of water so it is free to align due to the turning effect of Earth's magnetic field. Reproduced in William Gilbert's 1600 work *De Magnete*.

Georg Hartmann, a vicar in Nuremberg, appears to have been close to discovering inclination in 1544 when, in a letter describing declination measurements, he noted a slight tilting of his unweighted, pivoted needle. However, the angle of nine degrees that he recorded was much too low for his location.

Credit for finally recognizing inclination therefore probably belongs to an English hydrographer named Robert Norman. In *The Newe Attractiue*, a pamphlet published in 1581, Norman described inclination as "a newe discouered secret and subtill propertie concernyng the Declinyng of the Needle." The discovery was a breakthrough for mariners and scientists alike. *The Newe Attractiue* was reprinted at least four times (in 1585, 1596, 1614 and 1720), and later included in *Rara Magnetica*, a collection of works on geomagnetism published by Gustav Hellmann in Berlin in 1897.

Norman had gone back to the old idea of a floating compass, but his version was three-dimensional. He stuck a needle through a piece of cork and carefully adjusted the size of the cork until he just achieved neutral buoyancy in a goblet of water: the cork floated beneath the surface, neither rising to the surface nor sinking to the bottom. Norman found that the needle, once magnetized, pointed northwards (give or take a few degrees due to declination, which was very small in England at the time), but inclined downwards at about seventy degrees below the horizontal.

Norman also described the construction of the first "dip needle," a magnetized needle pivoted at its center of mass so it swung in a vertical plane. When this needle was oriented north–south, it finally settled at an angle below the horizontal equal to the inclination.

By 1600 inclination had been measured at many locations around the globe and a pattern was emerging. At the equator the inclination was close to zero: the dip needle was almost horizontal. The inclination steadily increased in steepness with latitude, until the dip needle pointed vertically downwards at the north (magnetic) pole and vertically upwards at the south (magnetic) pole.

Three centuries had now passed since Petrus Peregrinus had written his *Epistola*, making the case for experimentation in science and reviving interest in magnetism. Europe had emerged from the Dark Ages and the social and scientific climate had changed dramatically. The Renaissance had revived a passion for creativity and learning in both arts and sciences. As well as the ancient universities of Bologna, Oxford and Paris, centers of study now flourished across Europe at places such as Cambridge, Salamanca, Lisbon, Modena and Padua.

Given the slowness of travel and communication, it was remarkable how much interaction was taking place between academics and scholars scattered across the continent. Several great minds had been at work changing the way the universe was perceived—or trying their hardest to do so. But unfortunately for many, and disastrously for some, the Reformation in religious thought and tolerance had lagged some way behind the cultural and social renaissance. In earlier days, the geocentric model of the solar system had suited the young Christian church, from which Roman Catholicism emerged to dominate medieval Europe. The perceived circular orbits of the moon, sun, planets and stars about a central Earth were regarded as evidence of divine Creation: questioning such perfection was deemed as heresy, to be stamped out at all cost.

The problem was that the orbits of the planets had long been known to deviate from perfect circles: most notable was the so-called "retrograde" or occasional eastward motion of Mars across the sky. The suggestion that such observations might be explained better if all the planets, including Earth, orbited the sun can be traced back to the fifth-century BC Greek mathematician and astronomer Pythagoras and his followers. The idea of a sun-centered system had resurfaced several times down the following centuries, but had always failed to gain general approval. The overwhelmingly accepted solution to the problem of irregular planetary motions had been to modify the geocentric model with Ptolemy's idea of epicycles: extra loops superimposed on perfectly circular planetary orbits about Earth.

The heliocentric theory was finally revived again by the Polish astronomer Nicolaus Copernicus in the early sixteenth century. Copernicus, born in 1473, was an official of the Catholic Church and did not dare publish his theory for fear of excommunication, or worse. It was only thanks to his friends that his manuscript

De Revolutionibus Orbium Coelestium (*On the Revolutions of the Celestial Spheres*) was eventually printed in 1543: it was presented to him on his deathbed, when he was arguably beyond reprisal. Even so, his friends had inserted a disclaimer stating that, whereas the sun-centered model provided an excellent and superior means of making accurate predictions of planetary positions, it was no more than a hypothetical model and did not represent the universe in reality.

The Italian mathematician and astronomer Galileo Galilei, born twenty-one years after Copernicus's death, was rather more outspoken and so fared less well with church authorities. Throughout his life he was issued with warnings to teach the classical theory, but he resisted and was eventually brought before the Holy Office of the Inquisition in 1633, found guilty of heresy, and ended his life in imprisonment, albeit in his home.

By the beginning of the seventeenth century, however, the Reformation was sweeping northwestern Europe, and with many countries becoming largely reformed there was a more clement environment for the publication and dissemination of new ideas. In 1600 William Gilbert, a fifty-six-year-old physician with a passionate interest in the natural sciences, was fortunate to find himself in the relative safety of a moderately reformed Elizabethan England, with access to the works of his predecessors and contemporaries, publishers eager to circulate new ideas, and an audience hungry for new scientific discoveries. The time was right for a milestone work, and Gilbert, having conducted many years of research on magnetism and electricity, was ideally placed to write it.

Magnus Magnes

Magnus magnes ipse est globus terrestris.

— WILLIAM GILBERT, 1600

William Gilbert was born in Colchester, England, the eldest of eleven children. Historians are divided as to whether he was born in 1540 or 1544, but agree that he entered St. John's College of Cambridge University in 1558 when he would have been either fourteen or eighteen. It was not unusual for talented children of well-to-do parents to begin university studies young, so this does not help with ascertaining his birth date. In any case, he gained a Bachelor of Arts in 1560, a Master of Arts in 1564, and finally became a Doctor of Medicine in 1569.

Gilbert then spent several years traveling around Europe, during which time he seems to have developed an interest in magnetism.

William Gilbert (*c.* 1544–1603), "electrician," physician to Queen Elizabeth I, and author of the epic work *De Magnete.*

On his return to London in 1573 he began practicing medicine and was elected a Fellow of the Royal College of Physicians, a society in which he would subsequently hold several offices, eventually becoming president in 1600. This was a prodigious year for him in several other ways as well: he was called to court and appointed physician to Queen Elizabeth I, and he published a monumental book on Earth's magnetism. It would become his most famous piece of work.

Until his move to court, Gilbert had hosted regular meetings at his home, and here small groups of amateur scientists would discuss the science and philosophy of the day and the investigations they were undertaking. Such groups were forerunners of the Royal Society, which would be formally founded later in the seventeenth century with the explicit aim of promoting and fostering the exchange of ideas between scientists. Early on, Gilbert's non-professional scientific interests had leaned towards chemistry, but he had soon switched to electricity and magnetism, and during his years as a physician he devoted his spare time and considerable personal funds to the experimental investigation of these subjects.

De Magnete, Magnetisque Corporibus, et de Magno Magnete Tellure (*On the Magnet, Magnetic Bodies,* and *the Great Magnet of the Earth*) is a formidable work. In a style reminiscent of the Bible, it is divided into six "books" and 115 "chapters." Fortunately for modern readers, some chapters are very short and their titles are descriptive, in some cases almost abstracts in their own right.

De Magnete fulfilled a number of distinct purposes. It summarized and evaluated existing knowledge and philosophy of magnetism, described Gilbert's numerous experiments, and was a treasure trove of his thoughts and conclusions, doubtless enriched through presentation and discussion at his house meetings. It has been described as the first textbook on the subject, but it is a good deal more. As late as 1822 a doctor, John Robson, would optimistically proclaim: "It contains almost everything that we know about magnetism."

De Magnete was written in Latin, and almost before the ink had dried there were cries for an English translation. However, even in English—for example, the 1893 translation by P. Fleury Mottelay—it is far from easy to digest. Nevertheless, few textbooks survive over 200 years as the seminal work on any subject: *De Magnete* was something special. Gilbert's work ranks alongside that of Galileo, and of Sir Isaac Newton later in the seventeenth century. Galileo himself received a copy of the book and was clearly impressed and influenced by it.

In Book I, Gilbert separates well-founded information about magnetism from speculation and myth. He endorses the importance of experimentation, and scorns metaphysical philosophy without practical verification as pointless:

> . . . but they wasted oil and labor, because not being practical . . . having made no magnetical experiments, they constructed certain ratiocinations on a basis of mere opinions and old-womanishly dreamt things that were not.

Leaving aside what would today be regarded as blatant ageism and sexism, Gilbert's message was unmistakable: from the seventeenth century on, scientific theory would be firmly grounded in experimentation. Although Peregrinus had carried out experiments three centuries years earlier, the practice did not take hold until Gilbert expounded its importance. The message was to be reiterated even more forcefully by Francis Bacon a few years later, whereupon it gained the title Baconism.

Not all of Gilbert's conclusions were taken on board immediately. Like Porta, for instance, he ridiculed the idea that a lodestone smeared with garlic would lose its "virtue," yet throughout the next century British sailors who were found with garlic on their breath still risked a flogging for fear they would demagnetize their ships' compasses.

In the final chapter of Book I, Gilbert presents his most famous conclusion: *"Magnus magnes ipse est globus terrestris"* ("The Earth itself is a great magnet"). The experimental justification for this, and much else, follows in the next four books, which set out to discuss magnetic movements, direction, declination (which Gilbert calls "variation") and inclination (which he calls "dip" or sometimes, confusingly, "declination").

The final chapter of the first book (along with Book VI) is a rare early foray into geophysics, containing the first modern-style, reasoned discussion of the internal composition and structure of the Earth. Gilbert notes that the water-filled ocean basins, although apparently of great depth, are insignificant compared with the huge size of the Earth. He marvels at the irregular and enormously varied nature of the planet's surface, criticizing the ancient Greek philosophers who lumped together the vast diversity of rocks, sediments and soils as just one element, "earth." He wonders at the origin of these diverse surface rocks and sediments and discusses possible compositions of the interior, lamenting man's inability to probe beyond the 500 fathoms of the deepest known mine. (One fathom is six feet, or 1.83 meters, so 500 fathoms is just over 900 meters. Even now the deepest hole drilled on land, the Kola Superdeep Borehole in the former USSR, reaches just over twelve kilometers into the crust, only one five-hundredth of Earth's radius.)

Gilbert also notes that nearly all rocks seem to show some degree of attraction towards a lodestone:

> . . . provided only they not be fouled by oozy and dank defilements like mud . . . or that a greasy slime oozes from them like marl . . . they are all attracted by the loadstone . . .

And he seems to wonder if magnetic attraction in Earth's interior may be enough to bind and keep the planet in its spherical form.

This was a significant step forward in scientific thinking: it would be another eighty-seven years before Isaac Newton published his theory of a universal force of gravitation, a fundamental force of nature that acted between each and every particle of matter irrespective of composition, and that we now know accounts for both the binding together of planetary bodies and their orbital motions. In 1600, Galileo had yet to risk his neck describing the apparent imperfections of Creation, and the German astronomer Johannes Kepler had yet to show that planetary orbits were not circular but elliptical in shape. It was hardly surprising, then, that Gilbert and others invoked magnetism to explain geophysical and astronomical observations.

In Books II to V of *De Magnete* Gilbert describes his many experiments, including those he had conducted on a lodestone sphere, or terrella, which had led him to conclude that the Earth itself was a great magnet. In experiments remarkably similar to those of Peregrinus, he had plotted the orientations assumed by "bits of iron wire, one barleycorn in length" placed at different locations on the surface of the terrella. (A barleycorn is an old Anglo-Saxon measure equal to one-third of an inch, or approximately 8.5 millimeters.)

Unlike Peregrinus, Gilbert knew all about the angle of inclination and how it increased with latitude, from zero near the equator to 90 degrees near the poles. He had, therefore, immediately recognized the pattern described by the barleycorn-length wires: the angles they made with the surface of the terrella duplicated exactly the inclinations of the dip needle at equivalent latitudes on Earth.

He reasoned that just as the magnetism of the terrella was due to the magnetic material of the lodestone, Earth's magnetism must originate in the magnetism of the planet itself—not beyond the heavens, as had been argued, for example, in 1545 by a Spanish

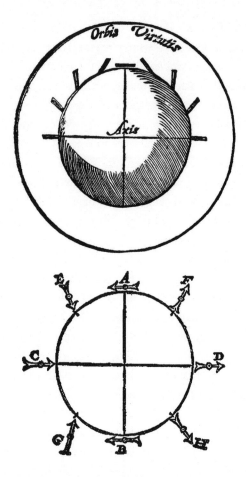

Sixteenth-century physician William Gilbert was deeply interested in
electricity and magnetism, and believed strongly in experimentation,
then a new scientific concept. *Top*: Gilbert's demonstration of the
orientations of short iron needles placed around a magnetized lodestone
terrella, or "little earth." The poles of the terrella are at opposite ends of
the "axis," where the needles stand at right angles to the surface.
Bottom: Gilbert's sketch of the inclination of a dip needle placed at
various latitudes around the Earth. Earth's north geographic pole is at
left (point C) and south pole at right (point D). Notice the resemblance
to the orientations of the iron needles around the lodestone terrella in
the illustration above.

geographer, Martinus Cortes, who, Gilbert wrote, "would be content with no cause whatever in the universal world;" not at the celestial poles, as favored by Peregrinus; not at the Pole Star; and certainly not in polar mountains of lodestone. Instead, Gilbert said, it was spread through the whole substance of the Earth.

Gilbert's concept of a uniformly magnetized planet, symmetrical about its axis of rotation, was a masterstroke. Nineteenth-century mathematicians would take the idea further, showing that the magnetic effect of a sphere of uniformly magnetized lodestone is identical to that of a bar magnet located at its center. Each has two poles, and so is commonly called a magnetic "dipole." Today Gilbert's concept is technically known as a *geocentric* (centered on the Earth) *axial* (along the axis) *dipole*.

But there was a complication with Gilbert's theory that magnetism was spread evenly through the Earth and that was declination. Gilbert was well aware of the phenomenon, since by now deviations of the compass needle from true north had been noted all over the explored world. If Earth's magnetism were truly identical to that of a terrella and symmetrical about the rotation axis, there would be no declination and the compass needle would point due north wherever it was placed on Earth.

By 1600 Gilbert had amassed a sizeable collection of declination observations. He knew that starting from the Azore Islands and moving east, declination increased only as far as Plymouth on the south coast of England, where it reached a maximum of 13°24′ E. Further east the declination decreased again; at Helmshud in Finmark (a region in the extreme north of Norway and about 30 degrees east of Plymouth), the compass needle again pointed due north.

Furthermore, Gilbert found that although the declination at Corvo, an island in the Azores, might be zero, the same was not true along the whole meridian of the Azores. Nor, in general, was the declination constant along any meridian:

. . . still along the entire meridian of the island of Corvo the compass does by no means point due north. Neither in the whole meridian of Plymouth at other places is the variation 13 degrees 24 minutes, nor in other parts of the meridian of Helmshud does the needle point to the true pole.

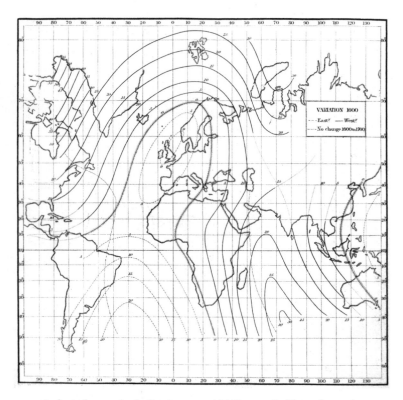

A chart of magnetic declination around 1600, compiled from observations that would have been known to William Gilbert. Compare this with Halley's chart of declination over the Atlantic Ocean a century later on page 58.

He therefore rejected Porta's assertion that declination was related simply to longitude as "vain hope . . . baseless theory . . . false as false can be."

Gilbert was unable to see any overall symmetry or regular pattern in his collection of declination measurements. He came to the conclusion that while the compass was influenced mainly by the magnetism of the globe as a whole—which he still believed to be symmetrical about the axis of rotation—superimposed on this was a secondary pull toward the magnetic rocks that made up the continents. This pull would, he thought, be strongest at the edge of the continents, so declination there would reach a maximum, as at Plymouth. It would decrease to zero in the middle of either the continents or the oceans, where the effects to east and west would cancel each other out, leaving the compass pointing due north.

Gilbert carried out a number of experiments designed to show this was a feasible explanation. In one, he gouged out an area on the surface of his terrella to represent the basin of the Atlantic Ocean. In another, he added lumpy blobs of lodestone to resemble the raised areas of continental land masses. He then repeated his observations, this time with two barleycorn-long iron wire "compasses." These behaved as he predicted: those in the middle of the "oceans" and "continents" pointed towards the pole of the terrella, while the others deviated in the direction of the simulated continents.

Gilbert went on to maintain that the continents were permanently fixed in position and so declination should remain constant at any given place on Earth:

> . . . forever unchanging, save there should be a great break-up of a continent . . . as of the region of Atlantis whereof Plato and ancient writers tell.

There was, of course, no way he could have foreseen the twentieth-century discovery of continental drift and plate tectonics.

Even so, it seems curiously out of character that this pragmatic man should have given credence to the legend of Atlantis. Whatever the case, Gilbert was soon to be proved wrong: the notion of a static and unchanging magnetic Earth would be gone forever.

Meanwhile, in March 1603 Elizabeth I died, presumably through no negligence on the part of Gilbert, who continued as physician to her successor, James I. Indeed, it is said that the queen left just one personal legacy, a bequest to enable Gilbert to continue his scientific experiments. Unfortunately, though, the great scientist had little chance to use it: he succumbed to the plague and died just eight months later.

Gilbert's explanation of declination meant the only way to get a complete picture was to go and make compass measurements over the entire globe. Until this was done, the compass and the angle of declination could provide little help in solving the longitude problem. By now, however, finding an answer to longitude had become something of a holy grail, and the idea that declination might provide a simple solution was too tempting to discard readily.

Hot on the heels of *De Magnete*, and not long before Gilbert's death, an eccentric French aristocrat, Guillaume de Nautonnier, Sieur de Castelfranc-sur-Lot, had thrown his hat into the ring. De Nautonnier, Geographer Royal in the court of Henri IV, was a mathematician and astronomer. Although he had an interest in Earth's magnetism, he evidently did not share Gilbert's dedication to the experimental method. Instead, he preferred to stay with the Greek tradition of geometric calculations based on pure hypothesis.

In 1602 he published his own mammoth tome, *Mécometrie de l'Eymant*, which, loosely translated, means *Determination of Longitude by Means of the Lodestone*. In it he claimed to have independently deduced that the Earth is a great magnet. However, without experimentation his claim lacked the validation and authority of *De Magnete*.

De Nautonnier took issue with Gilbert's explanation of declination, pointing out inconsistencies in his arguments. There was some truth in this. In Book I of *De Magnete* Gilbert had noted how insignificant the topography of the Earth was in comparison with the planet's size. This was true: as we now know, while Earth's radius is some 6400 kilometers, the highest mountains rise only ten kilometers and the deepest ocean trench lies only about ten kilometers below sea level. However, in Book IV Gilbert had used this same topography to account for quite significant angles of declination. Instead of developing a new explanation of declination, however, De Nautonnier tried to revive Mercator's idea of an offset between the magnetic and geographic poles, but now he argued that it was the magnetization of the whole Earth that was tilted with respect to the rotation axis. The idea was simple, but having a physical explanation—a uniformly magnetized Earth—did not improve the fit of the actual declination observations at all. De Nautonnier's model ran into all the same problems that Mercator had encountered in trying to calculate the locations of his magnetic poles. It was just *too* simple.

However, in 1602 de Nautonnier was still adamant that a simple declination pattern must exist. He was convinced that, once found, it would lead to a method of determining longitude and so he doggedly pursued the concept of a tilted magnetization. It is not difficult to see how tilting the magnetization and offsetting the magnetic poles produces a longitude-dependent declination pattern. Picture Gilbert's geocentric axial dipole, with its poles on Earth's axis of rotation. The situation is completely symmetrical: wherever a compass needle is placed on Earth, it will point due north towards the pole along a meridian, or line of longitude.

Now imagine tilting the magnetic axis so the magnetic poles move to opposite points in Siberia and Antarctica—de Nautonnier calculated the positions to be 67° N, 150° E and 67° S, 30° W. A compass needle will still point directly towards the magnetic north pole.

Imagine a new set of "magnetic meridians" outlining the alignments of compass needles over the globe—a network of great

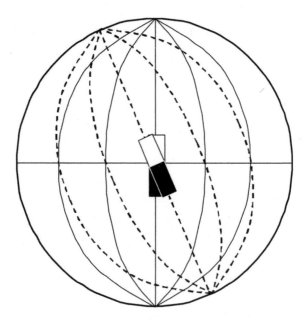

At any given point on Earth's surface a compass needle points in the direction of the magnetic meridian. The solid line depicts meridians due to a geocentric axial dipole—essentially a magnet at the center of Earth aligned with the rotation axis—as proposed by William Gilbert. The dotted line depicts meridians resulting from a tilted geocentric dipole, as suggested in 1602 by Guillaume de Nautonnier, a geographer in the court of French king Henri IV, to account for observations of declination and as a possible means of determining longitude. The angle between a magnetic meridian and true north is equal to the declination.

circles connecting the south magnetic pole to the north magnetic pole. These new magnetic meridians form a symmetrical pattern about the new magnetic axis. Two—the ones that pass through the geographic poles as well as the magnetic poles—coincide with geographic meridians, and along these the compass needle points due north. In other words, along these meridians the declination is zero (or 180 degrees along the short segments between the geographic and magnetic poles). In de Nautonnier's model these meridians are at longitudes of 330° E and 150° E.

All other magnetic meridians cut through the geographic meridians at angles that depend on location. These angles correspond to declination. Imagine taking a trip eastwards around the equator starting from 330° E. The declination will start at zero and steadily increase. After you have traveled through 90 degrees of longitude, it will reach a maximum 23 degrees at longitude 60° E. It will then decrease back to zero at 150° E, the longitude of de Nautonnier's north magnetic pole. When you move north or south from the equator, declination will increase with latitude, eventually becoming close to 180 degrees at locations between the geographic and magnetic poles.

Most of de Nautonnier's massive book was taken up with calculations and tabulations from which a mariner, armed with knowledge of his latitude and a measurement of declination, should have been able to work out his longitude. The latitude could be obtained by the traditional astronomical methods. Alternatively, a measurement of magnetic inclination could be used to calculate magnetic latitude, which itself is a function of latitude, longitude and the tilt of de Nautonnier's dipole.

All this was without doubt a very clever piece of mathematics, and it would have been an elegant solution to the longitude problem —if only the premise on which it was founded, the tilted geocentric dipole, had been completely accurate. Even today the tilted dipole is a reasonable approximation to the most significant features of Earth's magnetism but, like Gilbert, de Nautonnier was unwittingly up against the "non-dipole" component of Earth's magnetism—the small fraction that does not fit the symmetrical pattern of a dipole, whatever its orientation.

It would be twenty-seven years before an Italian monk would finally demonstrate experimentally that Gilbert's explanation of declination did not work either.

The Jesuits, a Roman Catholic order founded in 1534, had always placed a high value on education and scholarship. Although they maintained a traditional belief in an immobile Earth at the center of a perfect Creation, they were not averse to scientific experimentation to demonstrate the inner workings of Creation.

Niccolò Cabeo, a monk in the order, was interested in Gilbert's work on magnetism, and in 1629 he published a critique under the title *Philosophia Magnetica*. Supposing Gilbert's terrella to be about ten centimeters in diameter, Cabeo estimated the elevation of the continents and the depths of the oceans—constructed to scale—to be no more than one-tenth of a millimeter. He then showed that the effect of such tiny relief on the compass was negligible.

A new explanation was needed for the irregular pattern of declination over the globe. Although he had proved that Gilbert's suggestion of the magnetization of continental rocks was insufficient, Cabeo was more interested in defending Aristotelian metaphysics than in explaining irregular features of Creation. It would be some considerable time before anyone came up with a better explanation of this "non-dipole" component of Earth's magnetism.

Interestingly, Cabeo may never have seen *De Magnete*'s Book VI: it had been removed from certain editions destined for Catholic Europe. In it Gilbert had eventually turned his mind to the question of Earth's place in the universe and what role magnetism might play. He saw little option but to side with Copernicus. The distance to the stars must, he reasoned, be enormous. Therefore, to explain their apparent daily revolution around an immobile Earth, the stars above the equator had to be going at unimaginably high speeds. At the same time, those close to the pole—the Pole Star for example—must be hardly moving at all.

This was, Gilbert argued, highly improbable: "there cannot be diurnal motion of infinity, or of an infinite body." Instead, the Earth must be simply spinning on its axis beneath, or rather within, the whole universe of planets and stars, completing one rotation each day. As to what drove and maintained this rotation, Gilbert supposed it to be an interaction between Earth's magnetism and some sort of universal magnetic background.

In Catholic Europe, Gilbert's explanation would certainly have been considered heresy. In 1600, the year that *De Magnete* was published, an Italian philosopher and former priest called Giordano Bruno was burned at the stake in Rome for holding similar beliefs. Fortunately for Gilbert, England was more liberal. Nevertheless it was still deemed prudent to spare continental Europeans from the doctrine of Book VI.

The Wandering Compass Needle

*There is yet a further difficultie . . . great changes in the
Needles direction within this last Century of years, not
only at London, but almost all over the Globe of Earth . . .
the effect of a great and permanent motion.*

—EDMOND HALLEY, 1683

In 1634 Europe finally caught up with the secret that had lain
hidden for centuries in the layout of Chinese village streets—that
Earth's magnetism was not static and unvarying, but declination, in
particular, changed with time. In 1580 Gilbert's colleague William
Borough had measured the declination in London as 11°20′ E.
In 1622, however, Edmund Gunter, professor of astronomy
at London's recently founded Gresham College, recorded only
6°0′. Gunter was surprised by this decrease of more than five degrees
in scarcely forty years and, failing to come up with an explanation,
doubted the accuracy of both his and Borough's measurements.
However, in 1634 Henry Gellibrand, Gunter's successor at Gresham

College, recorded 4°6' E. It was clear that in London the declination was steadily decreasing.

Subsequent records would show this "secular variation" continuing. By the end of the seventeenth century a compass needle in London would point seven degrees west of north, while in the southern part of North America the declination would be to the east of north. In the space of one hundred years the pattern of declination across the Atlantic Ocean would change beyond recognition, completely invalidating Gilbert's theory that declination arises because the compass needle is attracted towards extra magnetic material in the continents.

Scarcely fifty years after Gilbert's death the challenge of understanding Earth's magnetism attracted the attention of another young Englishman, Edmond Halley. Although he is now remembered principally for the comet that bears his name, Halley would make long-lasting contributions not only to astronomy but also to geomagnetism.

The scientific environment had changed a great deal since Gilbert's time. Born in 1656, Halley was a contemporary of Isaac Newton, Newton's friend Christopher Wren and his adversary Robert Hooke, all early members of the Royal Society. Founded in 1660, and initially based at Gresham College, itself an independent institution for the promotion of learning, the "Royal Society of London for Improving Natural Knowledge" set out to enhance communication and collaboration between scientists. It had built on Gilbert's practice of regular meetings at which scientists could present and share ideas, solicited support for the work of its members and, sometimes controversially, fended off or sought to mediate in disputes of a scientific nature.

The society's principal publication, *Philosophical Transactions*, first appeared in 1663 and continues in an unbroken series to this day. It was one of the first scientific journals to be made available on the Internet, with all issues back to the first now accessible at the click of a mouse. Similar societies soon appeared in other major cities across Europe, where scientists such as Descartes, Leibniz and Huygens were at work.

More people, albeit almost exclusively men, were taking an active interest in science as it became clear that, more than just an academic curiosity, science had valuable practical applications. Scientific language and techniques were developing rapidly. Terminology was becoming more precise, enabling scientists to formulate their questions more clearly. Experimental methods and instrumentation were improving, and new mathematical tools, such as calculus, that naturally lend themselves to well-formulated physical problems, were emerging.

Edmond Halley was the son of a wealthy London merchant. His passion for both astronomy and geomagnetism were evident from his early days at the prestigious St. Paul's School. In 1672, at the age of sixteen, he measured the magnetic declination in London as 2°30′ W, and noted that the compass needle now pointed slightly to the west of north, not east of north as it had earlier in the century. Secular variation was continuing to draw the compass needle further towards the west.

From St. Paul's, Halley went up to Oxford, but in 1676 he cut short his studies when he was given the opportunity to join an astronomical expedition to St. Helena and the Ascension Islands, the goal of which was to map the stars of the southern hemisphere skies. As was to become the case in all his travels, Halley took his compass with him, and he made frequent measurements of declination. On his return, following the publication of his southern hemisphere star chart, Halley was elected a Fellow of the Royal Society, and at the bequest of the king was awarded the degree of M.A. by Oxford University. His scientific career was assured.

If 1600 had been a year of achievement for Gilbert, 1682 was an *annus mirabilis* for Halley. In August astronomers noticed a bright comet close to the sun. Halley studied its course closely and deduced that it must be the same comet as that seen in 1531 and again in 1607. Its elliptical orbit about the sun had, in the intervening years, taken it to the far reaches of the solar system. He calculated the period of the comet's orbit as a little over seventy-six years, and predicted that its next appearance would be in 1759.

Halley's Comet, as it is known today, appeared late in 1758, brightened and passed closest to the sun in March 1759. It has appeared on cue ever since. Once Halley realized that the comet's orbit might be slightly modified by the gravitational pulls of the planets, he correlated it with sightings in 1066 (famous for its coincidence with the Battle of Hastings), 1145, 1301 and 1456.

Halley was still only twenty-seven, and despite his diversion into comet research had not lost interest in geomagnetism. He had, in fact, been busy searching for a rational way to explain declination and its secular variation—in his words "to reconcile the observations by some general rule." He was not deterred by the pessimism of the prominent French mathematician René Descartes who, like Gilbert, apparently attributed declination to iron mines and lodestone deposits, randomly distributed about the Earth without any meaningful pattern.

Halley was looking for a solution with some degree of global symmetry, and "after a great many close thoughts" he came up with the idea of not two but four magnetic poles, two in the Arctic and two in Antarctica. He carefully estimated the locations of his poles. The European North Pole was "near the Meridian of the Lands end of England and not above seven degrees from the Pole Arctick." The American North Pole was "in a meridian passing about the middle of California, and about fifteen degrees from the North Pole of the World." The American South Pole was, he estimated, "about sixteen degrees" from "the South Pole of the World" and "some twenty degrees to the westward of Magellan's Straights," and the Asian South Pole "a little less than twenty degrees distant" from the south geographic pole and "in a Meridian which passes through Hollandia Nova and the Island Celebes." These corresponded approximately to 83° N, 354° E; 75° N, 241° E; 74° S, 270° E; and 70° S, 120° E, respectively.

Halley's reasoning was based on the assumption that a compass needle would be influenced most by the pole that lay closest to it; for example, the declination in Europe and the North Sea area would be dominated by the attraction of the European pole. He also supposed that secular variation was the result of the gradual drift of the poles. This drift, he deduced, affected all four poles to varying degrees, but was generally westwards.

However, Halley was not going to overstate his case. He concluded his 1682 paper by listing three obstacles to the further development of this concept. First, he said, although in the "frigid" polar regions it was easy to see which pole was closest and therefore dominated the declination, this was not the case in the "torrid" latitudes nearer the equator. Here, declination was likely to result from the combined effect of more than one pole, and to calculate declination accurately you needed to know how the magnetic attraction decreased with distance from each pole.

Halley was well aware that the gravitational force between two bodies depends inversely on the square of the distance between them: for example, if you double the distance between them, the force is reduced to a quarter of its original strength. In fact, both he and Robert Hooke had independently discovered this some time before learning that Newton had beaten them to it. Halley subsequently cajoled Newton into writing up this and much more of his work on the movements of the planets, and in 1687 personally saw through the publication of Newton's great work, *Philosophiae Naturalis Principia Mathematica.*

Halley realized that progress in geomagnetism depended on a comparable mathematical understanding of magnetic force. He tried some experiments of his own, but without success. He then apparently urged Newton to investigate, but Newton does not seem to have been much inspired by magnetism. There are only a few very brief mentions of it in *Principia*, including a reference to "some crude experiments" that suggested magnetic force might decrease not with the inverse square of distance, as is the case with gravity, but more probably with the inverse cube of distance (in which case doubling the separation would decrease the force by a factor of eight).

The second barrier to further development of his concept, Halley noted, was that even the sparse and meager observations of secular variation then available painted too complex a picture to be explained by simple westward drift of the poles: maybe their latitudes also changed. It would, he concluded, take hundreds of years to accumulate sufficient observations "to establish a compleat doctrine of the magnetical system," and until such time the motions of the magnetic poles would remain "secrets as yet utterly unknown to Mankind." Whether this was visionary foresight, or a ploy to gain support for his future endeavors to chart declination over large areas of the globe, is anyone's guess.

Halley continued to wonder at the origin of Earth's magnetism, and in 1692 he came up with a novel explanation of his four poles. His idea may have been inspired by Gilbert's comparison of the Earth with a lodestone terrella, but it also bore uncanny similarities with modern concepts of the Earth's interior, as well as some features that nowadays seem quite absurd. In essence, Halley proposed that the Earth was composed of a solid inner sphere and a spherical outer shell, with some sort of fluid filling the space between them. Both the inner sphere and the outer shell were magnetized, and each carried two poles: the inner sphere carried the European North Pole and the American South Pole, while the shell carried the American North Pole and the Asian South Pole. Neither pair of poles was truly antipodal—that is, diametrically opposite each other—but this does not seem to have worried Halley.

To achieve secular variation, Halley allowed both the inner sphere and the outer shell to rotate about the geographic axis but at different rates: because of the viscous resistance of the intervening fluid, the sphere lagged further and further behind the shell. He estimated that it would take 700 years for the inner sphere to fall one whole rotation behind the shell, and so complete one cycle of secular variation.

Halley concluded with a curious discussion of the possible purposes of his shell and inner sphere in the overall scheme of Creation. He speculated that there might be several more shells, nested inside one another, corresponding in size to the planets Mercury, Mars and Venus and each carrying a pair of poles. He probably envisaged this as a means of accommodating yet-to-be-discovered complexities in the geomagnetic field, but he discussed at length the likelihood of life on the inner shells.

The Canadian geophysicist Michael Evans has described Halley's model as a sort of global apartment block. Halley, he suggests, may have been having a hard time with the church, and may have even missed out on a chair at Oxford on suspicion that he believed Earth to be eternal, rather than divinely created. This curious discourse may have been his defence. Whatever the case, in 1703 Halley eventually became Savilian professor of geometry at Oxford University, a position he would hold until his death in 1742, and in 1720 he succeeded John Flamsteed to become the second British Astronomer Royal.

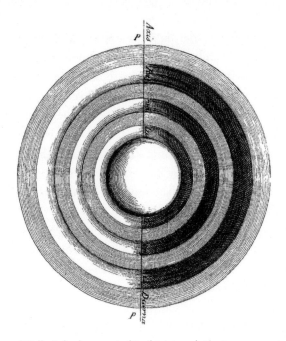

Edmond Halley's final concept of Earth's internal structure as a central sphere the size of the planet Mercury, and concentric shells equal in size to Mars and Venus. This was an elaboration of his earlier model, in which an outer shell enclosed a single inner sphere, with each carrying a pair of magnetic poles.

Before this, though, Halley scored another first for science when in 1698, in what has been claimed to be the first specially commissioned nautical geophysical expedition, he set out to measure and chart the magnetic declination (or "variation" as it was still called) over the Atlantic Ocean. Halley had been adamant that the route to understanding geomagnetic declination and secular variation, and possibly to determining longitude at sea, lay in collecting and documenting as much information as possible. He had made his case to the king, William III, and to the British Admiralty and been awarded a captain's commission.

On October 20, 1698 he set sail aboard the *Paramour* to, as the commission described it, "improve the knowledge of the longitude and the variations of the compasse" and to explore the south Atlantic further than previous mariners "till you discover the coast of Terra Incognita." The idea of this huge southern continent had existed since the ancient Greeks first speculated on the shape of the Earth and the distribution of its land masses. Most atlases and globes showed a *Terra Incognita* or *Terra Australis*, although the voyages of Ferdinand Magellan and Abel Tasman had already cast doubt on its extent.

Halley's first voyage was short-lived. Leaks in the ship and disputes with his second in command forced him to return from the West Indies after just a few months at sea. Problems fixed and lessons learned, Halley set sail again in 1699 and, making declination measurements all the way, he eventually reached latitude 52°40′ S on February 1, 1700. The closest thing he saw to a Terra Incognita were three huge icebergs that floated past on their way north, prompting him to name this part of the South Atlantic "The Icey Sea." The myth of a vast southern continent would finally be laid to rest when Captain James Cook circumnavigated New Zealand and Australia in the 1770s, proving the extent of the southern oceans.

In all, Halley collected over 200 measurements of declination, covering the length and breadth of the Atlantic. His next important task was to report to his sponsors. What was the best way to present his measurements? Although Halley's main interest was in Earth's magnetism, the justification for his work was to improve navigation, and his results would be used mainly by mariners and sailors. He needed to create a chart that would be readily understood and easy to use on the high seas.

His answer was to invent the contour line. Although he had made measurements only at discrete, disconnected locations along the route of the *Paramour*, Halley drew lines that joined up, in the simplest and smoothest way possible, places with equal declination values. These lines were drawn at intervals of one degree of declination, producing a series of equally spaced "Halleyan lines," which later came to be known as contours. Today we are familiar with the use of contours of equal height on topographic maps, and weather maps showing isobars or contours of equal atmospheric pressure, but in 1701 Halley's curved lines of equal declination were a revolutionary idea.

As was customary at the time, Halley's chart was adorned with elaborate text borders, compass roses and pictures. His route was marked out with tiny icons of the *Paramour*, while in The Icey Sea he depicted and described:

> . . . two sorts of Animalls of a Middle Species between a Bird and a Fish, having necks like Swans and swimming with their whole Bodies always underwater only putting up their long necks for Air.

Was Halley seeing penguins for the first time, or some obscure species never seen before or since?

In 1702 Edmond Halley produced a declination chart for the whole world using observations made by other mariners in other oceans. Surprisingly, though, he had never measured inclination. Although the accuracy of the dip needle was certainly inferior to that of the mariner's compass, the combination of declination and inclination measurements might have offered an opportunity for determining absolute location at sea and a solution to the longitude problem.

The first inclination chart, which covered the southeast of England, seems to have been made in 1721 by an Anglican priest called William Whiston. In addition to his religious calling,

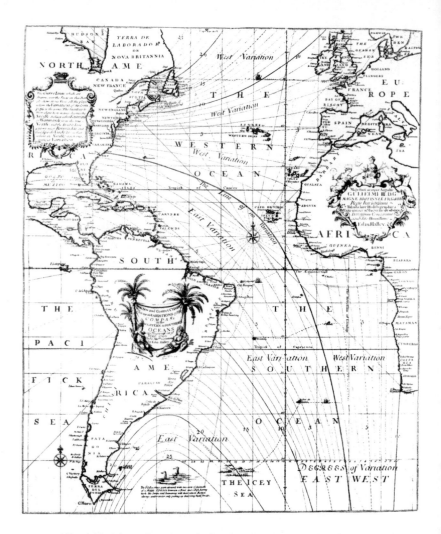

The earliest chart of magnetic declination, or "variation," over the Atlantic Ocean, published by Edmond Halley in 1701 from over 200 measurements made between 1697 and 1701. This was the first use of Halleyan lines, later known as contours. These lines of equal declination were at the time considered revolutionary.

Whiston was a keen mathematician and astronomer. At about the time Halley finally gained his Oxford chair, Whiston became Lucasian professor of mathematics at Cambridge, a position that had earlier been held by Newton, and would 200 years later be held by Stephen Hawking. However, he seems to have been too outspoken for the times, and in 1710 he lost the chair for apparently suggesting that Noah's flood might have had natural causes. Even in the liberal society of William III's England, academics had not achieved unbridled freedom of speech.

Like others before him, Whiston was tempted by the grail of solving the longitude problem. He had previously concocted several ingenious but unworkable schemes with his mathematician friend Humphrey Ditton; one involved regularly timed cannon blasts from a series of ships anchored at known locations throughout the oceans of the world. Now, on studying Halley's declination charts, Whiston realized that simultaneous measurements of declination and inclination might provide the answer.

The idea had considerable merit, but it was still compromised by the difficulty at sea of making sufficiently accurate measurements, particularly of inclination, because of the motion of the ship and iron objects near the compass. It would not be until 1768, just five years before the longitude prize was finally awarded to John Harrison for his fifth improved marine chronometer, that the first world inclination chart was finally published by Johannes Wilcke, a Swedish physicist.

Like declination, the angle of inclination at any given location was also soon found to undergo steady changes from year to year. Gellibrand's secular variation was slow, and could be detected only by collating and comparing compass and dip needle observations year after year. There was no indication yet as to where secular variation would lead in the long term. Between the times of Gilbert and Halley, the declination in London had shifted from 11° east to 7° west of north, a change of eighteen degrees. It was tempting to think the compass needle would soon turn again, and would eventually be found to slowly swing back and forth around true north, but there was little sign of this happening yet.

Then, just as geomagnetists and mariners alike began to settle into a routine of repeat observations, patiently amassing the data that might eventually lead to an interpretation of secular variation, another chance discovery occurred to upset their theories.

George Graham was a London-based maker of precision instruments. He made clocks, watches, quadrants and some of the most stable and sensitive compasses available, so in 1722 he was concerned to find that the needles of his best compasses were actually in constant, seemingly random motion. He went back and checked meticulously and found that:

> . . . all the needles I made use of would not only vary in their direction on different days, but at different times on the same day
> . . .

Sometimes the needles fluctuated by more than half a degree in a day.

To be sure he was not imagining things, Graham set up three compasses. Each had a needle 31 centimeters long, filed to a fine edge at the end. These needles moved over a scale that was graduated at intervals of one-sixth of a degree, or ten minutes. Subdivisions of two minutes could be estimated with the use of a magnifying glass.

All three compasses behaved the same way: Graham had discovered the daily variation of Earth's magnetism. This was quite different from the secular variation—at least ten times smaller in amplitude and something like 10,000 or 100,000 times more rapid. Further investigations showed that the average amplitude of these daily variations was greater in summer than in winter: in London it was about 13 minutes in summer, as opposed to seven in winter.

In Uppsala, Sweden, at a latitude of 60° N and 18 degrees further east than London, the astronomer Anders Celsius and his assistant, Olof Hiorter, had also noticed and become interested in the daily fluctuations of the compass needle. They coordinated their measurements with Graham, and between them discovered two features. First, the regular daily variation correlated with local time, beginning shortly after sunrise and quietening down with sunset. Secondly, on certain days the needle seemed to go crazy. On April 5, 1741, for example, Hiorter reported that "the needle at 2 PM began to be disturbed, so that at 5 PM it was 1°40' to the west of its declination at 10 AM," while Graham recorded:

> The alterations that day were greater than I have ever met with before . . . The observations on the other days were made with the same care, but they were much less and more regular.

These irregular disturbances were truly simultaneous at London and Uppsala, and were not a function of local time.

Following a hint Halley had given twenty years earlier, Hiorter and Celsius hit on the explanation: at the times when the compass needle was most disturbed, the northern lights, or aurora borealis, were putting on their most spectacular show. The two scientists wrote:

> A motion of the magnetic needle has been found that deserves the attention and wonder of everyone. Who could have thought that the northern lights . . . when they draw southwards . . . could within a few minutes cause considerable oscillations of the magnetic needle through whole degrees?

Gilbert's explanation of declination may have been disproved, but his chief argument—that the major source of Earth's magnetism lay within the Earth itself—had seemed incontrovertible. However, these new observations seemed to again point to a source outside the Earth: the daily movement of the compass needle was obviously related to local time, beginning at sunrise and ending at sunset, while the bigger disturbances that were simultaneous in London and Uppsala clearly correlated with auroral activity. Hiorter reasoned that the auroras must occur extremely high in the atmosphere. Here, then, were magnetic variations seemingly equally associated with processes in the heavens.

Was the source of Earth's magnetism internal or external, or were there perhaps two sources? How could this apparent dilemma be reconciled?

Measuring the Force

Numbers may be said to rule the whole world of quantity . . . the four rules of arithmetic may be regarded as the complete equipment of the mathematician.

—James Clerk Maxwell

The ancient Greeks had recognized that while lodestone attracted only iron objects, the translucent, golden, fossil-resin amber, when rubbed with fur, would magically pick up light objects of many different materials. Curiously though, while lodestone, the compass and Earth's magnetism had continued to intrigue both scientists and explorers, by the early seventeenth century the attractive properties of amber, or "elektron," as the Greeks had called it, had been almost forgotten. Even though Gilbert's portrait in the frontispiece of *De Magnete* was entitled "William Gilbert, M.D.—Electrician," the book was essentially about magnetism: Gilbert had devoted just a single chapter to electric attraction, the amber effect.

William Gilbert's versorium, a kind of compass consisting of a light needle balanced on a pivot. It was sensitive to nearby objects that had become electrically charged through being rubbed—for example with fur.

Gilbert had designed a sort of electric compass he called a versorium. It consisted of a light metal needle finely balanced on a pivot, and it was much more sensitive to the weak forces of electric attraction than were scraps of paper or chaff. With it, Gilbert found many more materials that displayed the amber effect. He compiled an impressive list of gemstones, including diamonds, sapphires, opals, amethysts and jets, as well as false gems made of glass or crystal, and used the term "electric" to describe any material that, after being rubbed, attracted the versorium's needle. Materials that did not do so he called "non-electric."

What caused this effect? Gilbert believed that an "electric" object, when rubbed, emitted an invisible cloud. This cloud, he supposed, filled the space around the electric material, giving it the ability to attract other objects. He called it an "effluvium" after the Latin *effluere*, meaning to flow out.

Strangely, he had shunned the idea of effluvia to explain magnetic forces, mainly on the grounds that a lodestone was able to attract a piece of iron even when solid objects were in the way. In other words, magnetic force could penetrate matter, whereas an effluvium would presumably be stopped short by it.

Despite his insistence on experimental verification in almost every other situation, Gilbert's explanation of magnetism remained esoteric. While avoiding the question of exactly how the magnetic force was transmitted through space, he maintained that magnetism was intrinsic to the substance of a lodestone, or a terrella. He had used this to argue that the source of Earth's magnetism lay within the planet. As he put it:

> The rays of magnetic force are dispersed in a circle in all directions; and the center of this sphere is not in the pole, but in the center of the stone and of the terrella.

Despite his familiarity with both attractive and repulsive forces in magnetism, Gilbert seems to have missed the phenomenon of electric repulsion. Once again this discovery was left to his successor, the Italian Jesuit Niccolò Cabeo. In the course of his many experiments, Cabeo noticed that a rubbed amber rod would first attract scraps of paper or the needle of a versorium, but as soon as the paper or the versorium needle had touched the amber, they would fly away—repelled in a similar way to two magnetic poles of the same kind. Gilbert's effluvium theory, which accounted only for electric *attraction*, was in deep trouble.

During the next century almost no progress would be made towards understanding electricity and electrical phenomena. In 1720 Willem Jacob 's Gravesande, in an early survey of physics, *Physices Elementa Mathematica, Experimentis Confirmata* (*Mathematical Elements of Natural Philosophy, Confirmed by Experiments*), was still describing electricity as:

> that property of bodies by which . . . (when rubbed) they attract and repel lighter bodies at a sensible distance.

Then in 1731, out of the blue, came a chance discovery that would propel electricity to the forefront of physics. Stephen Gray, the man responsible, came from Canterbury in England. Born in 1666, he had trained as a cloth-dyer but had developed an interest in astronomy, becoming a friend and colleague of John Flamsteed, the first Astronomer Royal, and working for a time as an astronomer. However, by the time he began experimenting with electricity he was destitute and living on a charitable pension at the London Charterhouse, a home for penniless gentlemen, retired soldiers and former servants of the king or queen.

Gray's experiments seem to have been designed more by trial and error than any logical reasoning, but nonetheless he discovered that he could transfer the electrical ability to attract chaff and scraps of paper from one end of a long string or cable to the other, provided the string was made of one of Gilbert's "non-electric" materials—a metal wire, for example—and was suspended or supported only by "electric" materials. Gray had discovered electrical conduction.

Gray's experiments eventually outgrew his rooms at the Charterhouse and he moved them into the paddocks of a mansion house in Kent, where he finally succeeded in conducting electricity through a 245-meter-long line. Gray had no real idea just what was being conducted from one end of his line to the other, and so he continued to experiment in a haphazard manner, testing the electrical conductivity of all sorts of materials from water to live chickens and human beings.

By the mid eighteenth century, electricity was in danger of becoming more of a fairground sideshow than a scientific endeavor. An event planned as a serious public lecture would often become oversubscribed by an audience attracted by the promise of a charged atmosphere, sparkling displays and maybe a shock or

Stephen Gray's demonstration of the conduction of electric charge through a human body—that of a Charterhouse "charity boy." The boy was suspended from silk threads, and his legs charged by friction. The picture shows scraps of chaff that have been attracted to his chest, his left hand, and to a conducting ball held in his right hand.

two. However, the spin-off of such frivolity was publicity, and with publicity came growing interest and, in time, financial support for further research. Electricity and magnetism were now set to move to center stage, not only in science but also in industrial and social development.

Gray's discovery of conduction spawned a new generation of theories of electricity. Like contemporaneous ideas about heat, these entailed hypothetical fluids that were supposedly transferred between materials on rubbing—at this time the only known means of electrifying an object.

In France, Charles-François Du Fay, a multi-talented scientist and the superintendent of Le Jardin du Roi (the Royal Botanic Gardens), and his former student and colleague the Abbé Jean-Antoine Nollet concocted a theory involving two such fluids. Normally an object or material was supposed to contain equal amounts of each fluid, evenly mixed. When rubbed with fur, an amber rod, they said, gained an excess of the "resinous" fluid, while a glass rod when rubbed with silk acquired an excess of the other, "vitreous" fluid. Two bodies carrying excesses of the same type of fluid repelled one another, while ones carrying excesses of different fluids would attract. Add to this the ability of the fluids to conduct through or between non-electric materials, and all the known electrical phenomena of the day could be explained.

At much the same time, and probably quite independently, a young American named Benjamin Franklin was beginning to take an interest in electricity. Franklin's later life of political activism, which would lead to his involvement in the Revolutionary War and drafting of the Declaration of Independence, would take him away from science, but not before he had left a lasting mark: the nomenclature of "positive" and "negative" charges, and the direction of "conventional" current trace their origins to Franklin and his theory of electricity.

Franklin's "one-fluid" concept was part-way between a fluid theory and a theory of electric particles. His fluid, he reported, consisted of "extremely subtle" electrical particles which repelled one another, but which were attracted to equally subtle particles of normal matter. When a glass rod was rubbed with silk, it became "charged" with an excess, or "positive," amount of fluid. On the other hand, when an amber rod was rubbed with fur, fluid was transferred out of the rod, leaving it with a deficit, or "negative" charge. Franklin's positive charge therefore corresponded to an excess of Du Fay's vitreous fluid, while his negative charge corresponded to an excess of the resinous fluid.

Franklin's concept would prove long-lasting, but his terminology of "positive" and "negative" would have an unfortunate impact on the way electrical currents are described to this day. Imagine bringing a normal (uncharged or neutral) conductor into contact with a positively charged glass rod—one carrying an excess of Franklin's fluid. Some of the excess will flow from the rod on to the conductor—that is, from positive to normal (or neutral). If the same conductor is touched to a negatively charged amber rod, which has a deficit of Franklin's fluid, fluid will flow from the conductor on to the rod in an attempt to remedy the deficit. This time, fluid flows from normal (uncharged or neutral) to negative.

Physicists have chosen to define the direction of "conventional" current as the direction in which Franklin's conceptual fluid would have moved, namely from positive to neutral, neutral to negative. This is to the perpetual consternation of today's students, who accept that electric current is, in fact, the flow of negatively charged electrons, which must, therefore, be moving in the opposite direction.

Research into both electricity and magnetism had now reached a stage where further progress could be achieved only by making quantitative measurements. In his *Principia*, published in 1687, Isaac Newton had put the science of mechanics on a firm mathematical footing, such that if some properties of a system were known, others could be calculated or predicted.

For example, Newton had defined the notion of force in a way that meant it could be used to calculate its effect on the motion of a body. A force acting on a body would give the body an acceleration that depended directly on the strength of the force: the bigger the force, the bigger the acceleration. Using this, Newton had shown that the elliptical orbits of the planets around the sun—by then thoroughly investigated by Kepler and Galileo, and almost universally accepted—could be accounted for only if there were a force of attraction between the sun and each planet that decreased as the distance between them increased. To be exact, the force was inversely proportional to the square of the separation.

He had argued that this previously unrecognized force of nature—the universal force of gravity—acted between each and every particle or object in the universe according to their masses.

Newton's mathematical equation for the gravitational force contained one other factor—the universal gravitational constant, G. Had Newton been able to directly measure the force between two objects, he could have determined this constant. However, the size of the gravitational force between two everyday masses is so tiny compared with the force that the Earth exerts on each—their weight—that the task was impossible with the equipment Newton had available to him. However, he did note that once the constant was known it would be possible to estimate the mass and density of the Earth from the weight of any object. (Another century would pass before this was finally achieved by another Cambridge physicist, Henry Cavendish.)

At Halley's instigation, Newton had also begun some experiments on the force between magnetic poles and had seemed to expect a result similar to the force of gravity. However, the task was complicated by the impossibility of obtaining single magnetic poles. The so-called "broken magnet paradox" was well known in Newton's day: if you break a magnet in two, rather than producing separate north and south poles you get two new magnets, each with north and south poles of their own. To this day, scientists have failed to isolate, or even find evidence of the existence of, magnetic "monopoles." Probably frustrated by this, Newton did not pursue the matter of magnetic forces, but moved on to other studies.

Several people now began to wonder if electric forces and magnetic forces could lend themselves to mathematical description. One such person was a Frenchman, Charles-Augustin de Coulomb. Coulomb had been born in 1736 in Angoulême. From an early age he had been fascinated by mathematics and astronomy, and by the age of twenty-one had already presented several papers to the Société Royale des Sciences in Montpellier. Although his mother had planned a career in medicine for him, the meticulous and analytical Coulomb had rebelled and trained as a military engineer. His scientific accomplishments would be sandwiched into a professional life as an engineer and architect of numerous military buildings throughout France and its offshore islands.

Coulomb had initially been attracted to study magnetism by a competition advertised by the Paris Académie des Sciences in 1773 and again in 1775. Inspired by George Graham's observations of the daily fluctuations of the compass needle, the Académie sought to find "the best manner of constructing magnetic needles, of suspending them, of making sure they are in the true magnetic meridian, and finally of accounting for their regular diurnal variations."

Coulomb's interest in measuring the mechanical strength of materials, particularly those used in building, had already led him to construct and perfect an extremely sensitive "torsion balance." The principle of the instrument was quite simple. Originally, Coulomb had been interested in how a thread or wire resisted being twisted, so he had suspended a rod of some sort horizontally on the end of a long thread or wire, all inside a closed glass container. When a controlled force was applied to one end of the rod in a horizontal direction at right angles to the length of the rod, the whole system twisted, until the torsional resistance of the suspending wire counteracted and balanced the applied force. At this point, the angle through which the rod had rotated could be read from a scale mounted around the wall of the container. This gave a measure of the torsional strength of the suspending thread: a torsionally strong thread would rotate through only a small angle, while a weaker thread would twist through a much bigger angle.

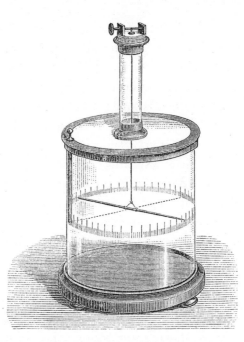

A sketch of Coulomb's torsion balance, which he used to investigate the force between the poles of two magnets. A magnetized needle was suspended horizontally from a silk thread, and a long magnet was inserted vertically through a hole near the edge of the lid, so that its lower pole approached and repelled one end of the needle. The angle through which the needle twisted depended on the force between the pole of the magnet and the needle.

Coulomb realized he was now faced with a slightly different situation: the competition required an instrument that would respond to extremely small variations in the force that Earth's magnetism exerted on a compass needle. If he were to replace the rod of his torsion balance with a long magnetized needle and the wire with a single "silk thread drawn from a cocoon," he would have just that instrument.

Coulomb's new suspended compass won him a share of the Académie's prize, but it did much more. It became the basis of a generation of instruments that would be the mainstay of magnetic observatories, surveys and laboratories right up until the advent of electronics and computers.

For five months before submitting his entry, Coulomb took a series of measurements several times a day. In general, he found that the declination increased during the morning, reaching a maximum around one o'clock in the afternoon. It then decreased more slowly until early evening, and remained quite steady overnight. Coulomb agreed with others—including Hiorter and Celsius—that since this regular activity was concentrated during the daytime, its cause was likely to be associated with the sun.

Some scientists had suggested that, since a magnet lost its magnetization when heated, the heat of the sun was gradually demagnetizing the Earth. Coulomb disagreed, arguing that if this were the case Earth's magnetism would long since have vanished altogether. Already interested in the effects magnets had on each other, he speculated that the sun was also one of Gilbert's great astronomical magnets, and "acts on the terrestrial globe as a magnet acts on another magnet." He also observed, like Graham and Celsius, that, from time to time, the daily variations became large and erratic, and noted that such disturbances were usually followed by spectacular night-time auroras.

Coulomb's compulsively analytical streak meant he was not satisfied just to record the daily variations of geomagnetic declination. Interesting as the tiny wanderings of the compass needle were, he was also fascinated in the fundamental nature of electricity and magnetism. Coulomb now set himself the task of finding a simple mathematical description of magnetic and electrostatic forces.

First, he needed to modify his apparatus to measure the extremely weak forces between electrically charged bodies, and between the poles of magnets. Details of his modifications, experimental methods and results, published in *Mémoires de l'Académie Royale des Sciences* in 1785, showed that Coulomb was eventually able to measure forces as small as one ten-thousandth of a grain of barley. (The grain, based on the mass of a grain of barley, was the basic unit of mass in the British imperial system, and equivalent to about one-fifteenth of today's gram.)

For his electrostatic experiments, Coulomb made the rod of his torsion balance of a light insulating material, with a pith ball at one end and a counterweight at the other. The pith ball could be charged, and the force between it and a second charged ball then measured over a range of different separations. A screw knob at the top of the suspension controlled the separation of the two balls.

In his first series of experiments, Coulomb showed that whether the two pith balls carried unlike charges and attracted each other, or carried like charges and so repelled each other, the force between them decreased as they got further apart. If their separation were doubled, the force decreased to a quarter of what it had been. If the separation were trebled, the force was reduced to one-ninth of the original. In mathematical terms, the measured forces seemed to depend on the inverse square of the separation, just like Newton's force of gravity. And just as Newton's gravitational force depended on mass, Coulomb found that as he increased the amount of charge—or, in his words, the "electric mass"—on his pith balls, the electrostatic force between them increased proportionately.

When it came to magnetism, Coulomb quickly recognized the problem of identifying the exact positions of the poles of a magnet, and the difficulty he would have isolating the effect of one pole while eliminating the effect of the other. He noted however:

> We have seen that the magnetic fluid in our steel wire twenty-five inches long was concentrated at its end in a length of two or three inches; that the center of action of each half of this needle was about ten lines from its ends . . .

Put another way, although it was impossible to isolate a single pole, the poles of a long magnet were very close to its ends. (Ten lines was about ten-twelfths of an inch, or just over two centimeters.) Coulomb hoped that by using a very long magnetized needle or rod to exert a force on the needle of his torsion balance, the effect of one pole would dominate, and the other pole would be so far away that its force would be negligible. By this method he was able to show that both the force of repulsion between like poles and the attractive force between unlike poles also decreased with the inverse of the square of the distance between their "centers of action."

He rounded off by stating that "the magnetic fluid acts by attraction or repulsion in a ratio compounded directly of the density of the fluid," although it is not clear how, or even whether, he actually demonstrated this last part.

Coulomb was not finished yet. Being a consummate perfectionist, he could not stop at deriving such important relationships by just one method. What if there was some elusive fundamental flaw in his torsion balance technique? Just in case, he went on to verify all his results again using a quite different kind of experiment, one based on timing oscillations. If an ordinary compass needle is displaced slightly from north and then released, it will swing back and forth at a rate that depends on the strength of the forces tending to restore it to north—in this case, the forces of Earth's magnetism on the poles of the needle. The stronger the force, the faster are the oscillations. In fact, the square of the frequency—that is, the number of oscillations in one second—is proportional to the force. Using suspended pith balls and magnetized needles similar to those in his electric and magnetic torsion balances, Coulomb reverified his inverse square laws.

All three known fundamental forces of nature—gravity, electrostatic and magnetic—had now been shown to follow an inverse square law. Like Newton's law of gravitation, Coulomb's laws of electrostatics and magnetism included elusive constant factors. The determination of these would have to await further developments in the theory of electricity and magnetism.

Lastly, Coulomb turned to the question the origin of magnetism, and came up with a theory of what happens inside a magnetic material that is still pertinent today.

Coulomb was well aware of the broken magnet paradox—that if you break a magnet in two you make two smaller magnets, each with north and south poles; do the same again and again and again, and while the magnets get smaller and smaller, each will always have two poles, north and south. He was also aware of a growing acceptance in the scientific community that, at the microscopic level, matter was made up of huge numbers of tiny indivisible particles, or molecules. He reasoned that the magnetic fluid, if it existed at all, was confined within individual molecules of magnetic materials such as iron. It could move *within* a molecule, he theorized, giving it a north pole and a south pole and so turning it into a mini-magnet. But it could not move *between* molecules. Inside a piece of ordinary unmagnetized iron, the north and south poles of the molecular mini-magnets would normally attract and abut one another, their effects would cancel each other out, and so there would be no overall magnetic effect. On the other hand, when a needle became magnetized—for example, by stroking it with another magnet—the molecular mini-magnets would become aligned, north pole to south pole, all along the needle, so at the ends there would be free north and south poles: the poles of the now magnetized needle.

Coulomb's work brought to a close the phase in the study of electricity and magnetism that had been concerned only with static charges and poles. Coulomb himself would survive the French

Revolution by retiring to the countryside. During this period two sons were born, and apparently some time later he married their mother, who was some thirty years his junior. Around 1800 he was appointed to overhaul the by then public French education system, and he and his family returned to Paris. He died six years later at the age of seventy.

Meanwhile, the world was set for a rush of discoveries and inventions that would establish an intimate relationship between electricity and magnetism—and open up a new range of possible explanations of Earth's magnetism.

Of Forces and Fields

*Ten thousand years from now, there can be little
doubt that the most significant event of the nineteenth
century will be judged as Maxwell's discovery of the
laws of electrodynamics.*

RICHARD FEYNMAN, 1964

Once again, a stroke of pure luck would launch a new chapter in
the story of magnetism. One day in 1791 Luigi Galvani, a professor of anatomy at the University of Bologna, was routinely dissect-
ing a newly dead frog when suddenly and completely unexpectedly
the frog's leg twitched of its own accord. Galvani had touched a
nerve with one of his metal instruments.

Having checked that the frog was indeed dead, Galvani carefully
repeated the process. The same thing happened. After considerable
thought, he decided the twitching must be due to some sort of
electric effect in the frog's nervous system.

Galvani's report describing his discovery of "animal electricity" was read by a colleague at the University of Pavia, Alessandro Volta, who immediately set about conducting his own experiments. In one he found that by placing a piece of tin foil on top of his tongue with a silver coin underneath he experienced a peculiar acidic taste. Eventually he deduced it was the metals, rather than the animal tissue, that generated the electricity, and in 1800 this led him to construct his famous "voltaic pile": the world's first electric battery. The pile consisted of alternate layers of copper and zinc, each separated by a layer of cardboard soaked in a solution of brine.

For a short time after this it was customary to distinguish ordinary electricity, which was generated by the old traditional method of friction or rubbing, from this new voltaic, or chemical, electricity. Ordinary electricity could be stored in a Leyden jar, an early form of capacitor, but when the terminals of the jar were connected together with conducting wires, the jar discharged instantaneously, producing only a single burst of electricity. On the other hand, voltaic electricity could be produced continuously, creating a steady flow of electric charge—a "current."

One of the first applications of Volta's battery was the electrolysis of water by Englishman William Nicholson and German Johan Ritter in 1800. When a current was passed through water between electrodes of silver and zinc, bubbles of hydrogen and oxygen gas formed around the electrodes. Other applications of electric currents followed. However, no one could have foreseen the next amazing breakthrough.

There are various accounts as to how it came about. All agree that Hans Christian Ørsted, a Danish professor of physics and chemistry, was giving a demonstration to a group of students during a lecture at his home in Copenhagen. According to some accounts, he was trying to show his students that electricity and magnetism were unrelated phenomena. What happened, though, was that when Ørsted held a wire carrying an electric current over a compass, the compass needle swung around until it was at right angles to the wire. When the current was northward, the compass swung to point west. When the current was reversed, the compass needle also reversed: a southward current made it point east.

It seems Ørsted did not immediately grasp the importance of his discovery: he merely wrote about it, in Latin, in a private letter to a few friends and colleagues. But just a few months later, on September 4, 1820, French physicist François Arago announced Ørsted's discovery of the magnetic effect of an electric current to a meeting of the Paris Académie des Sciences. At a second meeting the following week, Arago gave an impressive demonstration of Ørsted's experiments, and several members of the audience immediately seized upon their true significance.

Among them was a gifted mathematician, André-Marie Ampère. Ampère's life had already been marked by spells of professional and scientific brilliance, interrupted by personal tragedies and periods of severe depression. He had mastered all known mathematics by the age of twelve, but at seventeen he had lost his father to the guillotine during the French Revolution and virtually abandoned his studies. A few years later, when his young wife became ill and died, he was consumed with guilt because his work had kept him away from her for much of their brief marriage. A second marriage would end in separation, and a subsequent affair also proved disastrous, destroying Ampère's relationships with his children, one from each marriage.

Arago's announcement of the magnetic effect of an electric current spurred Ampère into action. For the next few weeks, the forty-five-year-old genius scarcely slept, carrying out numerous experiments and presenting demonstrations at the Académie. By the end of the year he had announced a series of important results, as well as his own theory of what he called "electrodynamics."

Publication took a little longer: it was 1827 before his *Memoir on the Mathematical Theory of Electrodynamic Phenomena, Uniquely Deduced from Experience* appeared. It has since been praised as the *Principia* of electrodynamics—the first work to lay out an integrated mathematical theory of electricity and magnetism.

Ampère's reasoning went along these lines: a current-carrying wire had been shown to behave like a magnet; since two magnets attracted or repelled one another, depending on which poles were brought close together, two current-carrying wires should also attract or repel each other.

His experiments showed that two parallel wires carrying currents in the same direction did indeed attract each other, while parallel wires carrying opposite currents repelled each other. There was, though, a difference: whereas the effects of a magnet or lodestone were concentrated at their poles, Ampère found that with a current-carrying wire the magnetic force was distributed along its whole length.

In the last few months of 1820 two other Frenchmen, a physicist called Jean-Baptiste Biot and his assistant Félix Savart, were also frantically beavering away trying to understand this new phenomenon. Both they and Ampère reported regularly to the Paris Académie so it is unclear who deserves credit for the next discovery. The result is, nevertheless, known as the Biot-Savart Law.

The scientists pictured a current-carrying wire as many tiny segments joined end to end, with the current passing from one segment to the next all along the wire. They found that each such segment produced a magnetic force, and the strength of this decreased with distance from the segment—as the inverse square of the distance. In effect they had discovered yet another inverse square law of nature. The tiny segments did not, of course, exist in isolation: to predict the magnetic force due to the whole current-carrying wire, the scientists had to add together the contributing forces due to each—the mathematical process of integration. Permanent magnets such as lodestones, bar magnets and compass needles, each of which had two poles where the magnetic effect was concentrated, were clearly, then, not the only source of magnetism. A long current-carrying wire had no poles, but it exerted a magnetic force along its whole length. This would prove to be a critical discovery.

Ampère was convinced that the two sources of magnetism were not independent, but intrinsically one and the same. He now set about finding the missing link. First, he imagined a circular coil of wire carrying a current. Picturing this current loop as many tiny segments, and adding up the magnetic forces due to each segment, he used the Biot-Savart Law to show that such a loop had exactly the same magnetic effect as a short bar magnet. In particular, when the current was clockwise it was as if a north pole lay above the loop and a south pole below it. When the current was counterclockwise, the opposite was true. Like a bar magnet, a current loop had two poles: it too was a magnetic dipole.

Ampère then reconsidered Coulomb's explanation of permanent magnetism. He concluded that at the microscopic level all magnetism must be "electrodynamic" in origin. Perhaps Coulomb's molecular mini-magnets were actually tiny current loops circulating in the atoms or molecules of magnetic materials.

Meanwhile, 300 kilometers away on the other side of the English Channel, a young scientist at London's Royal Institution was keenly following the fusion of electricity and magnetism into "electromagnetism." Although Michael Faraday's earlier work had been in the field of chemistry, he was about to make discoveries that would revolutionize not just physics but the day-to-day lives of the whole world. His experiments would directly result in the invention of

Michael Faraday, who was born near London in 1791, the son of a blacksmith. He made discoveries in electricity and magnetism that revolutionized not just physics but the everyday lives of people across the world, and led to the dynamo theory of Earth's magnetism.

the electric motor, the transformer and, most importantly for geomagnetism, the electric dynamo or generator—and provide an exciting new explanation of Earth's magnetism.

Faraday's story is a classic one: a poor boy, born and brought up on the streets of London, who grows up to develop an unstoppable passion for science, and achieves academic eminence through a career of hard work filled with brilliant discoveries.

Faraday was born in 1791 in Newington Butts near London. His father, James, was a blacksmith who, in search of work, had brought his young family south from Westmorland the previous year.

The family was deeply religious, and belonged to a little known Christian sect, the Sandemanians, which had broken away from the Presbyterian Church of Scotland. The Sandemanians strove to live according to the humble principles of Christianity laid out in the Bible, and these principles seem to have stayed with Faraday throughout his life.

James Faraday was often ill and there was rarely enough food to go around. There were no prestigious schools for Michael, who would later write that his education was "of the most ordinary description, consisting of little more than the rudiments of reading, writing and arithmetic at a common day school. My hours out of school were passed at home and on the streets."

In 1804, at the age of twelve, he left school and was sent to work as an errand boy for a bookbinder. After a year he became an apprentice, and by spending every spare moment reading the books brought in for binding he enhanced his meager education. Apparently these books included an early edition of *Encyclopedia Britannica*; the chapters on electricity and magnetism fascinated Faraday.

By chance, a customer who had noticed Faraday's interest in science offered him tickets for a series of lectures to be given by a famous chemist, Sir Humphry Davy, at the Royal Institution in Albemarle Street, Piccadilly. Faraday leapt at the opportunity, not only attending the lectures but scrupulously noting down every detail. Afterwards he neatly transcribed his notes, illustrated them with sketches, bound them and sent them to Davy, requesting consideration should a job arise at the Royal Institution.

Before long a vacancy did arise. Some sources say a fight at the Royal Institution's lecture theatre led to the dismissal of Davy's chemical assistant; others tell of a laboratory explosion in which Davy was temporarily blinded, the result being that he needed a secretary and note-taker. Whatever the truth, early in 1813 Davy apparently found himself in need of assistance, and called on the twenty-one-year-old Faraday.

The Royal Institution was, and still is, a unique scientific establishment. It was founded in 1799 by Sir Benjamin Thompson, also known as Count von Rumford, a colorful globe-trotting American whose name crops up all over science, most famously as the person who quashed the theory that heat was a fluid. (As a result of observing the boring of cannons in Austria, Thompson concluded that heat was, in fact, associated with molecular motion at the microscopic level. It was this that eventually led to the concepts of energy and the conservation of energy as it changes from one form to another.)

From the beginning the institution's goal was to promote a practical, applied approach to science. Its stated mission included "the diffusion of knowledge," "facilitating the general introduction of useful mechanical inventions" and "teaching . . . the application of science to the common purposes of life." Rumford had installed Humphry Davy as assistant, then lecturer, and finally as professor and director of the laboratory. It was here that Davy made numerous discoveries in chemistry and invented the miner's safety lamp that bears his name, but even he is reputed to have said that his best discovery of all was Faraday.

Almost straightaway on being hired, Faraday was taken on a grand tour of Europe. This was supposed to be an extended honeymoon for Davy and his new wife but the trio seem to have visited a great many scientists and taken in all the important scientific centers of the day. No wonder Mrs. Davy is reported to have been somewhat grumpy with her husband's new scientific assistant. Faraday, meanwhile, made the most of his opportunities to meet men such as Volta and Ampère. Both fostered his interest in electricity, so in 1820 he quickly grasped the importance of the news filtering through from Paris.

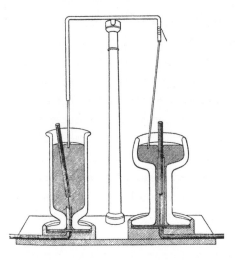

The apparatus used by Faraday to demonstrate the magnetic forces between magnetic poles and current-carrying wires. Both cups contained liquid mercury and a long magnet fixed at the base. Electric current entered each through the wire at the top. In the left-hand cup, the magnet was pivoted and its upper pole rotated in response to the magnetic force exerted on it by the fixed current-carrying wire. In the right-hand cup, the lower end of the current-carrying wire rotated in response to the force exerted on it by the magnet.

Faraday's first series of electromagnetic experiments concerned the directions of the magnetic forces exerted between magnets and current-carrying wires: he wanted to find out whether the magnetic effects of currents and magnets were really the same. He figured that the magnetic force produced by a current, as observed by Ørsted and Ampère, should cause a free magnetic pole—if one could be created, or at least approximated—to move in a circle around a current-carrying wire.

Like Coulomb, Faraday used a very long rod-shaped magnet so that its poles were far apart. He fixed the magnet at its lower end in such a way that it could tilt in any direction as its upper pole responded to the magnetic force around a current-carrying wire. A conducting circuit was created by carrying out the experiment in a cup of mercury, as shown on the left of the illustration opposite. Faraday found that the upper pole did indeed revolve in a circle around the end of the wire. He had built the first electric motor.

Imagining there should be reciprocal magnetic interactions between a current-carrying wire and a permanent magnet, he now investigated the opposite effect. If a current-carrying wire exerted a force on the pole of a magnet, then a permanent magnet should also exert a force on a current-carrying wire. In the right-hand cup of his apparatus he fixed the upper pole of a magnet, while leaving the end of a current-carrying wire free to move. As he had suspected, the wire moved in a circle around the pole of the magnet.

A further curious fact was that the force on the wire was not in the direction of the current, nor in the direction a free magnetic pole would be expected to move if placed at the location of the wire. Instead it was at right angles to both. This apparently dawned on Faraday on Christmas Day, 1821. Was his mind wandering during the Christmas sermon, or after enjoying Christmas dinner prepared by his new wife in the rooms they shared above the Royal Institution?

News of Faraday's discoveries spread, but his achievements were not received warmly in all quarters. His background still counted against him, and more eminent scientists envied his remarkable physical intuition. Over the next few years he would be accused of copying the work of others, and his nomination to become a Fellow of the Royal Society would be opposed by a number of influential scientists—including, surprisingly, Humphry Davy himself. However, he weathered the experience, was finally exonerated, and in 1823 was awarded a fellowship. On Humphry Davy's retirement two years later Faraday succeeded to the directorship of the Royal Institution. His career was now set to soar.

Faraday was convinced there was further reciprocity to be found in electromagnetism. Since an electric current was now known to produce a measurable magnetic effect, he reasoned it should be possible to produce an electric current magnetically—by using either a permanent magnet, or the magnetic effect of one current-carrying circuit on another, initially current-free, circuit.

He seems to have been temporarily sidetracked from this task by his collaboration with another young physicist, Charles Wheatstone, on vibrations related to sound waves. However, the diversion would prove to be serendipitous. The two men had observed that a metal plate set vibrating would induce vibrations in a similar nearby plate without actual physical contact. (We now know that the effect, resonance, is due to sound waves traveling through the air between the plates.) This led Faraday to hypothesize that similar invisible vibrations might explain the action-at-a-distance nature of the magnetic force, and be the means through which the electromagnetic induction he was seeking might take place.

If an electric current was, as he thought, a sort of wave motion involving tension and strains between charged particles in a conducting material, the space around a current-carrying wire, or a magnet, must be filled with magnetic "lines of force." Although these lines were invisible and impossible to detect, he believed they were real structures that exerted forces on magnetic objects. Each line followed the direction of the force that it exerted on a free north pole, while the density of the lines indicated the strength of the force. He was no doubt influenced by the familiar pattern traced out by iron filings or iron sand sprinkled around a magnet.

Faraday was not the only scientist struggling to achieve the electromagnetic induction of a current: the French had been given a head start. Like them and many others, he began by placing magnets close to circuits of conducting wires in which he had incorporated the most sensitive current meters or galvanometers. Again and again, though, this approach proved fruitless. Finally the penny dropped: an electric current was a stream of electric charge in motion. In his earlier experiment in 1821 it had been this moving charge that had exerted force on the free magnetic pole and made it revolve.

Faraday now deduced that to create a force on the charges in a conductor, and so induce a current in a conducting circuit, he needed to move a magnet in the vicinity of the circuit, rather than simply place one there and hope. Alternatively, and in accord with his resonance experiments with Wheatstone, if he were to change the pattern of magnetic lines of force in the vicinity of a circuit, this should have the same effect.

In August 1831 he finally struck gold. He later described his apparatus and experiment:

> Two hundred and three feet of copper wire in one length were coiled around a wooden block; another 203 feet of similar wire were interposed as a spiral between the turns of the first coil, and the metallic contact everywhere prevented by twine. One of these coils was connected with a galvanometer and the other with a battery.

Only at the instants of connecting or disconnecting the battery in the first circuit did Faraday notice a jerk (first one way and then the other) in the needle of the galvanometer in the other circuit; when a steady current flowed in the first coil, the galvanometer stubbornly recorded no current at all in the second. In other words, only during the brief periods when the current in the first coil was increasing and its magnetic lines of force growing, or the current was decreasing and the lines of force decaying away again, was a current induced in the second coil. For a current to be induced, the

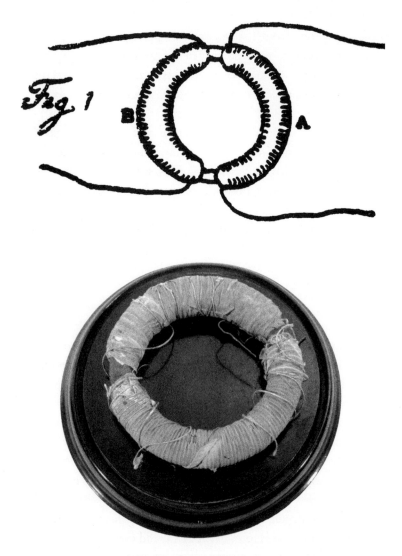

The apparatus through which Michael Faraday discovered the secret of electromagnetic induction. Two coils of wire are wound on the same wooden ring, but insulated from one another by twine.
Top: Sketch from Faraday's notebook. *Bottom*: Photograph of the actual coils, courtesy of the Royal Institution of Great Britain.

pattern of the lines of force around, and more importantly threading through, the second coil had to be changing.

Having discovered the secret of electromagnetic induction, Faraday now went on to experiment with permanent magnets. He found that moving a magnet into or out of a coil of conducting wire induced pulses of current in the coil. He also found that inserting a core of soft iron inside the two coils of his original experiment greatly enhanced the linkage between the lines of force, and so the induced current. He now had the elements of a basic transformer and all the ingredients of his famous law of electromagnetic induction.

Today, Faraday's law of electromagnetic induction is usually rendered in its mathematical form, something like this:

$$\varepsilon = -d\varnothing/dt$$

Astonishingly, though, Faraday did not write as much as a single equation in his laboratory notebooks. His genius was in conceptualizing physical phenomena and explaining them in words. He left the mathematics to his successors.

In a presentation to the Royal Society in November 1831, he also described the first working dynamo. This was designed to produce an electric current by rotating a copper disc in a magnetic field. In Faraday's original version, the disc had been rotated between the poles of a magnet, and a current had been produced between brush-contacts on the spindle and the edge of the disc. Good electrical contact had been ensured through the use of liquid mercury. He had subsequently found he could even produce a current using Earth's magnetic field instead of a permanent magnet.

Although Faraday's disc dynamo was conceptually simple, it was inefficient; before long more sophisticated dynamos would be designed for practical applications. But this was not the end. The

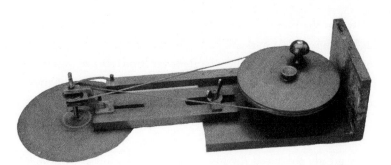

The Faraday disc dynamo. When the copper disc on the left is rotated between the poles of a magnet (not shown), a voltage is induced between its center and its rim. If a circuit is connected between the center and rim of the disc, the voltage will drive a current through it. This model was made at the University of Aberdeen around the time that James Clerk Maxwell was professor of natural philosophy there.

disc dynamo concept would reappear later, when the race to understand the magnetism of first the sun and then the Earth heated up.

Meanwhile, Faraday's scientific career, which he would spend entirely at the Royal Institution, forged ahead. He investigated the properties of electrical insulators and various magnetic materials, and the electromagnetic "polarization" of light, as well as formulating the laws of electrolysis. Last but not least, he gave memorable public lectures and scientific discourses. As well as his brilliant descriptions, he was renowned for the experiments with which he illustrated the lectures, bringing science to life for his audiences.

In 1826 he initiated two popular and enduring series: Friday Evening Discourses and Christmas Letures for a Juvenile Audience. Both continue to this day. (The Christmas Lectures are now broadcast on television.) In all, Faraday would personally deliver nineteen series of Christmas Lectures and 123 Discourses, and organize many more. One Friday evening early in 1846 he was caught short without a lecturer and had to step in at the last minute. The result was his "Thoughts on Ray Vibrations," published that May in *Philosophical Magazine*. In this lecture he speculated that the propagation of light, which by then was known to possess wave-like properties, might involve coordinated variations of his lines of magnetic and electric force.

Late in life Faraday became frustrated by a serious loss of memory. It has been suggested that this resulted from his liberal use of mercury to ensure good electrical contact in many of his experiments. Whatever the cause, in 1867 he resigned his directorship at the Royal Institution in favor of his good friend John Tyndall and declined a knighthood. On August 25 that year he died as plain Mr. Faraday.

For all his brilliance, Faraday's one limitation had been his lack of mathematical skills. Before long another scientist would emerge to fill this void and take knowledge of electromagnetism to a new level. James Clerk Maxwell hailed from Edinburgh, where he had been born into a well-to-do family in 1831. In a regrettably short professional life, which oscillated between Scotland and England, this gifted physicist would make pivotal advances in many areas of mathematical physics, from theories of color vision to determining how gases respond to changes in pressure and temperature. Most importantly, he would formalize the language of electromagnetism and lay down its mathematical foundations.

Much of Maxwell's early childhood was spent at his family's country estate, Middlebie, at Glenlair, about twenty miles west of Dumfries in Kirkcudbrightshire. Some time previously the Clerk family had inherited the estate from their relatives the Maxwells—hence the name Clerk Maxwell. From the age of about three James is said to have constantly asked, "Show me how it doos." Anything mechanical fascinated him, especially the house's plumbing system and the inner workings of the network of bells used to summon servants.

Maxwell's mother died when he was eight, and an aunt helped raise him. He attended the Edinburgh Academy, where it took several years for him to overcome his shyness and excel. At the age of thirteen he presented his first mathematical paper to the Royal Society of Edinburgh; its subject was the geometry of ellipses and related shapes. At sixteen he began to study mathematics, natural philosophy, chemistry and philosophy at Edinburgh University, and in 1851 he moved to Cambridge University, graduating in 1854. He narrowly missed out on the coveted title of "senior wrangler" awarded to the top student in mathematics, but shared the more prestigious Smith's Prize for an essay based on original research.

Becoming a Fellow of Trinity College, for the next two years Maxwell continued to study mathematics there while supervising undergraduate students. He might have stayed longer at Cambridge but his father became ill, and in 1856 he returned to Scotland to be closer to him. Before long he was offered, and accepted, the chair in natural philosophy at Marischal College, Aberdeen. He was just twenty-four. A year later he won the coveted Adams Prize offered by St. John's College Cambridge with a brilliant essay on the rings of Saturn, described by the respected mathematician, geophysicist and Astronomer Royal Sir George Airy as "one of the most remarkable applications [of mathematics] to astronomy that I have ever seen."

He married Katherine Dewar, the daughter of Marischal College's principal, but this did not help him in 1860 when Aberdeen University was restructured. Marischal merged with the other college, King's, and Maxwell, the junior of the two professors of natural philosophy, lost his job. In another blow, he narrowly missed out on a chair at Edinburgh University; *The Edinburgh Courier* hinted that despite his quick and brilliant mind and undoubted eminence in academic circles, his lecturing style may not have suited the average Scottish student:

Professor Maxwell is already acknowledged to be one of the most remarkable men known to the scientific world . . . There is another quality which is desirable in a professor in a university like ours and that is the power of oral exposition proceeding on the supposition of imperfect knowledge or even total ignorance on the part of pupils.

Fortunately, Maxwell did not remain unemployed for long. In 1860 he was grasped by King's College, London, and he and his wife moved south. The next five years would be his busiest and most productive, culminating in the 1865 publication of his paper "Dynamical Theory of the Electromagnetic Field" in *Philosophical Transactions of the Royal Society.*

Maxwell followed Faraday in thinking that a magnet or an electric charge somehow modified the space around it, so that when another magnet or charge entered that space it experienced a force. Faraday's lines of force had given a picture of the forces that could be expected at various points in space. Maxwell developed this idea by calling the region of altered space a magnetic or electric field. The field concept was to become central in many areas of theoretical physics, but Maxwell's original definition was deceptively simple. A field, he wrote, is "that part of space which surrounds bodies in electric or magnetic conditions."

This was just the start. Maxwell now tried to reconcile his fields with Faraday's "Thoughts on Ray Vibrations" and the famously enigmatic concept of "luminiferous aether" to which Isaac Newton had resorted when his corpuscle theory of light ran into problems. After Newton, a number of scientists —Christiaan Huygens, Thomas Young and Augustin-Jean Fresnel among them—had shown that light propagated more like a wave than a stream of particles, and the concept of "aether" had evolved to describe the medium through which these waves supposedly traveled. Faraday had observed that

James Clerk Maxwell, born 1831. Maxwell presented his first mathematical paper to the Royal Society of Edinburgh at the age of thirteen. By his early forties he had developed a mathematical theory to describe all known phenomena of electricity and magnetism. His four famous equations of electromagnetism are known simply as Maxwell's Equations.

a magnet could influence the way light traveled, and therefore considered that light itself might involve electric and magnetic vibrations. Maxwell reasoned that if the aether were the medium through which light traveled, it might also provide the support for Faraday's lines of force and his own magnetic field.

What, though, was this "aether?" Faraday had doubted it could be matter in the normal sense of the word. "Ponderable matter," he had written, "is not essential to the existence of physical lines of force." Nevertheless, Maxwell clung to the notion of a curious thin fluid that pervaded all of space, and even penetrated solid materials. He supposed it could be set in motion by electric currents and magnets, and that the energy of such motion could be transmitted through it by a series of exchanges between what we now call kinetic and potential forms of energy—an elastic wave. This was an important step: it enabled Maxwell to apply the physics of mechanics—in particular of mechanical waves—to problems of electromagnetism.

The aether would continue to prove elusive. Numerous experiments, most notably those of Americans Albert Michelson and Edward Morley in the 1880s, would fail to find any evidence for its existence, and in 1905 Einstein's theory of special relativity would finally remove the need to invoke it as anything more than a romantic notion. However, even when the mechanical scaffolding provided by the aether was dismantled, Maxwell's results still held true. Today, the field concept has not just survived, but become the backbone of many areas of theoretical physics.

Maxwell's next task was to formalize the language of electromagnetism. To do this, he had to define every property clearly and unambiguously, and lay down rules as to how each should be measured. Only then would his mathematics and equations work consistently.

First and foremost, he needed a more rigorous definition of "field," one that covered both the strength of the force that would be exerted on a charge or pole placed in the field and its direction. He defined the strength of an electric field as the magnitude of the force that would be exerted on a very small positive charge at a particular point in space, divided by the size of that charge. This imaginary test charge had to be very small so it did not change the overall field it was meant to measure. In the same way, the strength of a magnetic field could be defined in terms of the force on a tiny north pole. To fully understand the effect of a force it was also necessary to know the direction in which it acted. Push or pull? Up or down? To the left or right? Or somewhere in between?

Maxwell found that to rationalize and describe all the known phenomena of electricity and magnetism, he needed to define not just electric and magnetic fields, but a total of twenty different properties. His mathematical theory incorporated Coulomb's electrostatic and magnetostatic force laws, Ørsted's magnetic effect of a current-carrying wire, Ampère's results on the magnetic interaction between current-carrying wires, Faraday's force on a current-carrying wire in a magnetic field, and his law of electromagnetic induction.

Initially, he presented a set of twenty equations, but by 1873, when he published his *Treatise on Electricity and Magnetism*, he had distilled these to the four that have become universally known as Maxwell's Equations. Written in the shorthand of mathematical symbols, these equations look deceptively simple. They relate the variations of electric and magnetic fields both in time and space, and are expressed using the mathematical language known as vector calculus. They can be applied to electric and magnetic fields in any situation imaginable—from a single atom to the entire universe—including the deep interior regions of Earth.

Maxwell's equations predicted something else too, a wave motion consisting of coordinated oscillations of magnetic and electric fields, which travel through empty space at a speed of 300 million meters a second: Faraday's "ray vibrations." Maxwell readily acknowledged Faraday's prior insight:

> The electromagnetic theory of light, as proposed by [Faraday], is the same in substance as that which I have begun to develop in this paper, except that in 1846 there were no data to calculate the velocity of propagation.

According to Maxwell's analysis, the speed of his electromagnetic waves depended on just two parameters, electrical permeability and magnetic permittivity, which respectively expressed the electrical and magnetic properties of free space. In the twenty years since Faraday's "Thoughts on Ray Vibrations," these had been measured and so, unlike Faraday, Maxwell was able to show that his predicted electromagnetic waves traveled at a speed indistinguishable from the best available measurements of the speed of light.

Not twenty years after the publication of Maxwell's equations, a German physicist, Heinrich Hertz, would succeed in generating electromagnetic waves from oscillating electric currents in an antenna, confirming Maxwell's prediction.

Unfortunately, Maxwell was no longer alive to see this far-reaching endorsement of his work. In 1865, at only thirty-four, he had "retired" to his estate in the south of Scotland and embarked on writing *Treatise on Electricity and Magnetism*. During this period he had declined several offers of appointments, but in 1871 he accepted an invitation from Cambridge University to become the inaugural Cavendish professor of experimental physics; his first job was to design and supervise the building of the Cavendish Laboratory. After this he had set himself the task of sorting through and editing the mountains of unpublished paperwork left by the reclusive Henry Cavendish sixty years earlier. He discovered to his amazement that Cavendish had not only built a sensitive electrostatic torsion balance, but had also discovered the inverse square law of electrostatic force—fourteen years before Coulomb.

"Cavendish," he reported, "cared more for investigation than publication . . . it did not excite in him the desire to communicate the discovery to others." Today Cavendish is best known for his experimental determination of Newton's universal gravitational constant, work that did see the light of day during his lifetime, and that led to the first estimate of the mass of the Earth and its density.

Meanwhile, Maxwell's own research career was all but over. In 1879 he became ill and soon afterwards, like his mother, he died from cancer at the age of forty-eight.

The Third Element

*The instructions of the Royal Society, and the instru-
mental means prepared under its direction, provided for
the examination, in every branch of detail, of each of the
three elements which, taken in combination, represent,
not partially but completely, the whole of the magnetic
affections experienced at the surface of the globe* ...

—Edward Sabine, 1857

While the great nineteenth-century physicists had been laying the experimental and theoretical foundations of electromagnetism, other scientists continued to be fascinated by the intricacies of Earth's own magnetism. By the time Maxwell published his famous equations, knowledge of the variations and variability of Earth's magnetic field had increased beyond recognition, and ingenious new mathematical techniques had been invented to describe it in all its complexity.

By the end of the eighteenth century, mariners were routinely measuring and recording the angles of magnetic declination and inclination at locations throughout their voyages. Sometime during the 1790s the first attempts were made also to measure the third element of Earth's magnetic field: its strength, or intensity. Coulomb had shown that the strength of the force exerted by one magnet on another could be measured by timing the oscillations of a compass needle. Using a similar idea, mariners now began to time the oscillations of the dip needle with which they already measured the field's inclination.

The earliest measurements of the intensity of Earth's magnetic field may have been made on an ill-fated French expedition led by a naval captain called Jean-François de Galaup—or, as he styled himself, Comte de La Pérouse. In 1785 La Pérouse set sail from France in command of the aptly named frigates *l'Astrolabe* (the astrolabe) and *la Boussole* (the compass) on an ambitious scientific expedition to explore Alaska, Japan, Russia and the southern seas. A great admirer of Britain's Captain James Cook, who had taken scientists and artists on his expeditions, La Pérouse had with him ten scientists and illustrators. There had been keen competition to gain a place on the crew. A sixteen-year-old second lieutenant in the French military, Napoleon Bonaparte, had applied but been turned away. Had his application succeeded, the history of Europe may have been very different.

The La Pérouse expedition survived three years on the high seas, with members making many scientific observations, measurements and notes, before it arrived in Australia's Botany Bay in January 1788. After a six-week stay it sailed, planning to be back in France by June 1789. Alas, neither men nor ships were to be seen again.

In 1791 the French government sent two more ships, *la Recherche* and *l'Espérance*, to look for them. The ships were under the command of Joseph-Antoine Bruni d'Entrecasteaux, a noted explorer. He too took a formidable team of scientists with him. Although he failed to find any trace of La Pérouse's expedition, he managed to carry out a great deal of exploration and make

Drawing of the magnetic dip needle used on the 1791 Pacific expedition of French explorer Bruni d'Entrecasteaux, during which the first recorded measurements of magnetic intensity were made. D'Entrecasteaux died of scurvy, and the expedition was disbanded. When the ships' records were eventually returned to Paris, the magnetic measurements were published. This illustration appeared in the 1808 report of the expedition.

many scientific observations around Australia, particularly at Van Diemen's Land, today's Tasmania.

After leaving Australia, this expedition also ran into problems. The first were political. While journeying through the Pacific islands, the men picked up news of the ongoing revolution in France. While most officers on board were royalists, the crew was dominated by supporters of the revolution and trouble was never far from erupting. Then there was another crisis: the men were hit by scurvy, and d'Entrecasteaux himself died near the coast of New Guinea. Soon afterwards, in Java, the expedition was disbanded and the officers handed over the ships to the Dutch authorities there to prevent their falling into the hands of French revolutionaries.

Eventually Elisabeth Paul Edouard de Rossel, one of the most senior officers to survive, made it back to Paris. Here he was reunited with the ships' records, which included the earliest known measurements of geomagnetic intensity. In 1808 de Rossel published these measurements in *Voyage de D'Entrecasteaux, Envoyé à la Recherche de La Pérouse*, his two-volume transcription of d'Entrecasteaux's journals. The information was enlightening. It was clear that the higher the latitude, the faster the dip needle swung and the shorter its period of oscillation (that is, the time for a complete swing, one way and back to center, then the other way and back to center). In particular, the records showed periods of 1.869 seconds and 1.850 seconds in Tasmania in 1792 and 1793, while at locations near the equator the periods of oscillation were 2.404 and 2.429 seconds. Earth's magnetism was tugging more strongly on the dip needle in Tasmania, making it oscillate faster there than at lower latitudes. It was clear that the strength of Earth's magnetic field increased with latitude.

Unfortunately for de Rossel, because of various delays and crises he had been beaten to the publishers by an energetic Prussian explorer and geographer. Born in Berlin in 1769, Alexander von Humboldt had become interested in geology and mineralogy while studying at the University of Göttingen. This interest had taken him to the School of Mines in Saxony, and then to a government job supervising mining. One day, while surveying an outcrop of serpentine rock in the Fichtelgebirge, a mountainous region of Bavaria that had recently been acquired by Prussia, he noticed his compass behaving strangely. Each time he approached the rock the needle swung in a completely different direction: there seemed to be no sensible pattern to the directions it took up at different locations.

Von Humboldt had discovered with his compass what Magnus had found with his boots and staff several millennia earlier: the phenomenon of intensely magnetized rocks. In a moment of inspiration he guessed that these rocks, exposed on a mountain top, may have been struck by lightning, and that this might be the origin of their magnetization. Coming twenty years before Ørsted's discovery of the magnetic effect of an electric current, this was a remarkable insight.

The incident seems to have sparked in von Humboldt an enduring interest in Earth's magnetism. An inveterate traveler of independent means, in 1799 he set off from La Coruña in Spain on a scientific expedition to Central and South America in the company of Aimé Bonpland, a French botanist. (He had originally planned an expedition to Egypt followed by a circumnavigation of the world under the French flag, but Napoleon had put an end to that plan.)

During his five years in the Americas, Von Humboldt collected a multitude of data of all kinds, much of which he would eventually describe in his multi-volume treatise, *Kosmos*. He was one of the first people to speculate that South America and Africa may once have been joined together, and that the Atlantic Ocean was still in the process of widening. However, he considered the magnetic intensity measurements he made with his dip needle among his expedition's most important work.

With the equipment he had available, von Humboldt had been unable to measure the absolute value of the magnetic intensity; instead he had compared intensities from place to place. It was known that the period of oscillation of the dip needle depended inversely on the square root of the intensity. Put more simply, if the period of oscillation of the dip needle at site A was twice that at site B, the intensity of the magnetic field at A was one-quarter that at B. However, without knowing the actual intensity at B it was possible only to make comparisons of this sort.

Von Humboldt decided to set as a standard the intensity at a location near the equator, where the period of the dip needle was at its maximum value and intensity was, therefore, at its minimum. He chose the town of Micuipampa in northern Peru and defined the intensity there as exactly one unit. From his American expedition, together with his journeys around Europe, he accumulated about 120 measurements of both inclination and intensity (relative to that at Micuipampa) from locations ranging in latitude from 10° S to 52° N, and covering a 105-degree interval of longitude.

On his return to Paris in 1804, von Humboldt enlisted the help of the French physicist Jean-Baptiste Biot to publish his observations of geomagnetic intensity. Biot (the same man who, with Félix Savart, would later make fundamental discoveries about the magnetic effect of a current-carrying wire) was interested in Earth's magnetic field. With a chemist friend, Joseph Louis Gay-Lussac, he had made a pioneering ascent in a hot-air balloon to see whether magnetic intensity varied with altitude. The two men had hoped to solve once and for all the question of whether the magnetic field originated inside the Earth or outside it. If the origin were internal, the intensity should decrease as they moved further above the Earth's surface. If it were external, the intensity should increase. Unfortunately their dip needle had iced up, making measurements of its period of oscillation unreliable, and so their experiment had been in vain.

Von Humboldt and Biot now produced a chart that was divided into five "isodynamic zones," four in the northern hemisphere and one in the south. Each zone included a different range of geomagnetic intensity values, and the chart clearly showed a general increase in intensity from the equator towards the poles.

The first charts to show isodynamic contours, rather than just zones, would be published two decades later, in 1825 and 1827, by Christopher Hansteen, a physicist at Norway's University of Christiania (now Oslo). It had now also become customary to time the oscillations of a horizontally balanced or suspended compass needle rather than a dip needle, and hence measure the intensity of the horizontal component of the field. As Coulomb had found, by suspending the needle from a low torsion fiber—often silk—it was possible to circumvent the unavoidable friction of a mechanical pivot. This allowed the needle to swing for longer and gave a more accurate and precise result.

A good needle would swing for ten minutes or more before coming to rest. This allowed an observer to time several hundred oscillations. For his 1825 chart of horizontal intensity, Hansteen recorded the times for 300 oscillations of his horizontally suspended magnet at various locations. They included 753 seconds in Paris, 780 seconds in Oxford, 820 seconds in Edinburgh and 850 seconds in Bergen. The lengthening of the period of oscillation with increasing latitude indicated that, contrary to the total intensity, the horizontal component of the magnetic field decreased from equator to pole.

This was exactly what Hansteen had expected. He had recognized that the horizontal component of Earth's magnetic field at any location depended not only on the total intensity, but also on its inclination to the horizontal. At the equator the inclination was zero and the total field was horizontal, so the horizontal component was equal to the total field. At higher latitudes the inclination increased and the field steepened, until at the poles the total field was vertical and there was no horizontal component at all. He therefore knew from earlier measurements of total field intensity and inclination that the horizontal component should decrease from equator to pole. Conversely, having measured the horizontal component and the inclination, he could work backwards to calculate the total intensity. All things considered, this proved to be a more accurate way of obtaining total intensity than the old dip needle method.

In his 1827 chart Hansteen plotted the total geomagnetic intensity, still using von Humboldt's practice of referring the values to that at Micuipampa. As expected, the intensity increased steadily from equator to pole.

Hansteen was also intrigued by the morphology of Earth's magnetic field, its structure and form, and the origin of its secular variation. His interest had in fact begun when, as a student of Ørsted, he had seen some of Johannes Wilcke's charts of the southern polar regions. These were based on the declination measurements of James Cook and Tobias Furneaux, who had commanded the *Resolution* and *Adventure* respectively, during Cook's second voyage of discovery to Australia, New Zealand and the southern seas in 1772.

Hansteen had noticed that the polar magnetic field seemed to separate into two centers: a major one south of Van Diemen's Land and a smaller one south of Tierra del Fuego in South America. This observation of what appeared to be two south magnetic poles came to dominate Hansteen's view of terrestrial magnetism. In 1819, just a year before Ørsted's groundbreaking discovery and the advent of electromagnetism, he had published *Untersuchungen über den Magnetismus der Erde* (*Investigations of the Earth's Magnetism*) and with it won a competition set up by the Royal Danish Academy of Sciences in 1811, which posed the question: "In order to explain the magnetic phenomena of the Earth, is one magnetic axis sufficient or must we assume more?"

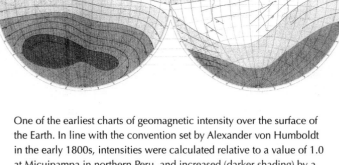

One of the earliest charts of geomagnetic intensity over the surface of the Earth. In line with the convention set by Alexander von Humboldt in the early 1800s, intensities were calculated relative to a value of 1.0 at Micuipampa in northern Peru, and increased (darker shading) by a factor of about two between the equator and the poles. This chart was compiled and published by Edward Sabine, a scientific adviser to the British Admiralty, in 1836.

This question, of course, went back more than a hundred years to Edmond Halley, who had suggested there were two pairs of magnetic poles: one pair in the Earth's rigid outer shell and the other fixed in an inner sphere. To explain secular variation, he had supposed that the inner sphere rotated more slowly, lagging further and further behind the shell, so that its poles appeared to drift slowly westward.

Halley's idea had received little attention at the time but now, over a century later, Hansteen had become an ardent advocate. First he compiled a 148-page collection of all the declination and inclination observations he could muster, many times more than had been available to Halley. He then examined them all for accuracy and consistency, and plotted charts of inclination for

1600, 1700 and 1780, and of declination for eight different dates between 1600 and 1800.

Hansteen argued that there was evidence for two "points of convergence," or focal centers, in the northern polar region, similar to those he had seen in Wilcke's south polar charts. At northern latitudes, the compass needle pointed towards two distinctly different focal centers: in North America it pointed towards a focus in Hudson's Bay, and in northern Europe to a focus in Siberia. He supposed that at other locations compasses were affected by both these focal centers, and so pointed towards neither one nor the other but in some intermediate direction.

Hansteen deduced that his four focal centers must correspond to Halley's four poles: the American and European north poles and the Asian and American south poles. Using observations made in the polar regions, mainly between 1769 and 1774, he carefully estimated their positions: 70°17′ N, 259°58′ E; 85°47′ N, 101°29′ E; 69°27′ S, 136°15′ E; and 77°17′ S, 236°43′ E.

Comparing these positions with Halley's from 1682, Hansteen concluded that all four poles were on the move. Both northern poles and the major southern pole (the Asian, or Australian) were moving eastwards around the geographic poles while the minor southern pole, the South American, was moving westwards. He even estimated how long it would take each magnetic pole to complete a circuit—4320, 1728, 1296 and 864 years, respectively—and suggested that these figures were of astronomical significance. Their lowest common multiple of 25,920 years corresponded, he claimed, to the period of precession of the equinoxes, precession being a variation in Earth's orbital motion that affects the cycle of seasons.

In the final part of his *Magnetismus der Erde*, Hansteen made a determined attempt to explain his observations, using the mathematics that Biot and a Swiss mathematician, Leonhard Euler, had

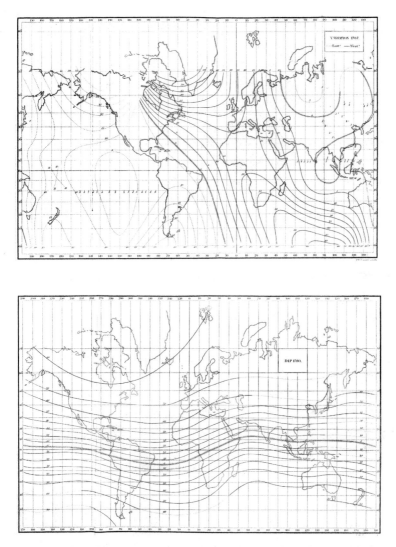

Top: Chart of geomagnetic declination (then called "variation") around the world, 1787. This was first published by Christopher Hansteen in *Magnetismus der Erde* in 1819, and later reproduced by Edward Sabine. *Bottom*: Chart of geomagnetic inclination ("dip"), around the world, 1780, first published by Christopher Hansteen in *Magnetismus der Erde* and later reproduced by Edward Sabine.

recently developed to describe the magnetic field of a dipole. With the discovery of electromagnetism still a year away, permanent magnetism remained the only known source of a magnetic field. A uniformly magnetized terrella and a bar magnet both had two poles, and both produced the same dipolar form of magnetic field. Gilbert had convincingly proposed that Earth's magnetic field was that of a geocentric axial dipole—that is, it was as though a dipole were located at the center of the Earth and aligned with the Earth's axis of rotation. Others such as Mercator and de Nautonnier had argued that a dipole tilted at a small angle to the rotation axis better matched the observed pattern of declinations and inclinations.

Gellibrand's discovery of secular variation, the slow changing of Earth's magnetic field, had led to the suggestion that the dipole axis was precessing—rotating with time—around the geographic axis. Now Hansteen was suggesting that if the observations were to be satisfactorily explained there had to be not one but two dipoles. As the theory of electromagnetism advanced, models of Earth's magnetism based on permanent magnets would become obsolete, but Hansteen made one last attempt to support the theory by proposing that these two dipoles were neither axial nor geocentric. By adjusting their positions within the Earth, their orientations and their strengths, he eventually produced a fit with which he was satisfied. It was, he claimed, "as well established, as a means of representing the phenomena, as any hypothesis whatsoever introduced in physical illustration."

By this time, however, even he was beginning to have doubts. His two-dipole model might fit the observations and be useful in forecasting secular variation, but in his final analysis he carefully avoided implying there actually were two dipoles within the Earth. Indeed, he did not even discuss whether the physical source of Earth's magnetism was internal or external.

With theories of geomagnetism based on permanent dipoles running into more and more difficulties, Ampère's discovery that a current loop produced exactly the same form of magnetic field as a dipole offered a new explanation. As early as 1831 Peter Barlow, an English mathematician and engineer, wrote a paper entitled "On the Probable Electric Origin of All the Phenomena of Terrestrial Magnetism." Barlow had constructed a wooden globe with copper wires wrapped around lines of latitude. On passing a current through the wires he had produced a magnetic field similar in form to the Earth's, and so demonstrated that Earth's magnetic field could be electrical in origin.

The question was how might electric currents be generated inside the Earth? Barlow suggested the heat of the sun was somehow transformed into electrical energy, but he did not manage to come up with a credible mechanism.

The Magnetic Crusade

The Earth speaks of its internal workings through the silent voice of the magnetic needle.

—Christopher Hansteen, 1819

Thanks to Ampère, Biot, Euler and others, mathematics was fast becoming a popular pursuit across Europe. Its language was evolving, and it was increasingly being used to construct accurate descriptions of all sorts of physical phenomena. One distinguished mathematician who would leave his mark on almost every area of physics, including the study of Earth's magnetism, was Carl Friedrich Gauss. Born into a poor family in Brunswick, Germany in 1777, Gauss had shown astounding mathematical abilities from a young age. At seven, while his classmates fumbled with chalk and slate, he had amazed his teachers by instantly coming up with the sum of the whole numbers from one to a hundred. The answer, he declared, was 5050: 50 times 101. There were 50 pairs of whole numbers, with each pair adding up to 101 (1+100, 2+99 and so on).

Gauss's father, Gerhard, worked variously as a gardener, bricklayer and canal tender, and assumed his son would also earn his living as a laborer. However, Gauss's mother Dorothea and his uncle Friedrich encouraged and fostered the child's interest in mathematics, and in 1792 the Duke of Brunswick granted the fifteen year old a stipend to attend the local Collegium Carolinum, today the Technische Universität Carolo-Wilhelmina zu Braunschweig (Technical University of Braunschweig). The young man was soon deep into the study of number theory; he derived the binomial theorem, a central result of number theory, and independently discovered what is now known as Bode's Law, a numerical relationship that relates the sizes of the orbits of successive planets.

In 1795 Gauss moved from Brunswick to the University of Göttingen, where his career and reputation flourished. In 1807, at the age of twenty-nine, he was appointed director of the astronomical observatory, and for the next ten years he worked mainly on mathematical problems in astronomy, using the now popular methods of calculus. He also perfected and published his "least squares" method for finding the best mathematical fit for a set of observational data marred by random errors, or "noise." The method proved itself in verifying the orbit of the asteroid Ceres, until then something of an enigma because of the difficulty of making accurate measurements of its position.

Before long, Gauss's interests had shifted to geodesy, the study of Earth's shape and dimensions, and potential theory, an emerging mathematical method that was found to be particularly useful for analyzing Earth's gravitational field. These would set the scene for his future work in geomagnetism.

Gauss was encouraged by von Humboldt. The two men first met in 1828, and it is likely that they discussed the problems

Carl Friedrich Gauss, born 1777. A mathematical genius from childhood, Gauss became fascinated by geomagnetism, producing the first mathematical model of Earth's magnetic field, building his own geomagnetic observatory, and setting up the Göttingen Magnetic Union, which aimed to establish a worldwide network of observatories.

involved in obtaining absolute measurements of magnetic intensity and mathematical representations of Earth's magnetic field. Also on the agenda would have been von Humboldt's vision of a worldwide network of geomagnetic observatories. Within the next few years Gauss had solved the absolute intensity problem, produced a mathematical model of Earth's magnetic field, built his own geomagnetic observatory, and thrown his weight into the geomagnetic observatory movement by setting up Göttingen Magnetische Verein (Göttingen Magnetic Union).

Gauss's answer to the intensity problem was as simple as it was ingenious. The obstacle to calibrating the period of oscillation of a dip needle or compass had been the need to know what amounted to the strength of the magnet used for the needle—in technical terms, its "dipole moment." The period of oscillation depended on the product of this dipole moment and Earth's magnetic field intensity—that is, the one multiplied by the other.

Gauss figured that by making a second measurement he could obtain the ratio of the two—the one divided by the other. He took another compass, and while letting Earth's magnetism pull its tip towards the north he arranged the original magnet to pull it at right angles—towards the east or west. Like a rotational tug-of-war, the angle at which the compass needle ended up depended on how strong the magnet was compared with Earth's magnetic intensity. Once both the product and the ratio of two things are known it is a straightforward matter of arithmetic to untangle them: Gauss was able to calculate not just the intensity of Earth's magnetism, but also the dipole moment of the magnet.

For obvious reasons this became known as the oscillation-deflection method. After Gauss published the details in 1832 it was rapidly adopted, not just in permanent observatories as he had intended, but also in some temporary observatories and in survey work. It also got around the problem of compass needles losing their magnetization during long sea voyages, something that had dogged explorers and surveyors for centuries and was the main reason ships routinely carried several compasses. Von Humboldt and others must have rued not having thought of such a simple solution.

Meanwhile, Gauss's keen mathematical brain had been racing ahead. If he could devise a simple mathematical description of Earth's magnetic field at any epoch of time—a standard set of equations from which, by inserting a few appropriate numbers or "coefficients," the declination, inclination and intensity at any location on Earth could be calculated—everyone everywhere would be able to compute whatever feature of the field they happened to need.

A group of French mathematicians, including Pierre Simon Laplace, Adrien-Marie Legendre and Siméon-Denis Poisson, had already made significant progress in developing such a method for describing Earth's field of gravity. This used the idea of gravitational potential (a simple intermediary quantity from which the strength and direction of the field could be calculated) and a technique they called "spherical harmonic analysis."

This technique capitalized on Earth's near spherical shape. The gravitational potential and force of gravity vary so little over Earth's surface that when you move from one place to another you cannot feel any difference. However, the Earth is not perfectly spherical. At the equator, for example, it bulges slightly and so locations on the equator are further from the center of the Earth than are the poles. This—and other, more complex, irregularities in the shape and the distribution of mass in the Earth—mean the gravitational potential and field are quite complicated. The essence of spherical harmonic analysis was the idea that this complex detail could be pictured as the combination of a series of smooth waves around lines of both latitude and longitude. These waves, or "spherical harmonic functions," involved a series of mathematical expressions (called Legendre polynomials), and the relative contribution of each was given by its own numerical coefficient. Adding together all the spherical harmonic functions, each multiplied by its coefficient, produced the mathematical model of the gravitational potential. From this the gravitational field could be calculated.

One of the most appealing features of spherical harmonic analysis was, and is, that each of the harmonic functions corresponds to a conceptual physical source. The first and simplest does not depend on latitude or longitude at all, and so represents a symmetrical, spherical mass. This makes by far the biggest contribution, and so dominates the mathematical model of Earth's gravitational field. The next function represents a dipole, the next a quadrupole and describes the "fat-in-the-middle, squashed-at-the-poles" effect of the equatorial bulge—and so on. You can choose to fit as many or as few spherical harmonic functions as you want, giving as simple or as detailed a picture as you want, or as the data will allow. Adding more "higher degree" harmonics simply adds more detail; it does not alter the contributions already calculated for the lower degree ones.

Gauss reasoned that with enough observational data it should be possible to carry out a similar spherical harmonic analysis of the Earth's magnetic field. Both the gravitational and geomagnetic fields involved action-at-a-distance forces that originated within the Earth. Both could be described via the intermediary idea of a potential. A mathematical model based on spherical harmonic functions would be more widely applicable, and might eventually prove more durable, than the physically appealing but ultimately unrealistic models based on permanent magnetic dipoles.

By 1838 Gauss was impatient to try out this idea. First, however, he needed up-to-date measurements of all three elements of the geomagnetic field—declination, inclination and intensity—from as many locations as possible and preferably spread evenly around the globe. By now declination and inclination charts were being produced and updated regularly. He chose Barlow's 1833 chart of declination, and an 1836 chart of inclination attributed to a cartographer by the name of Horner and published in the German *Physikalishe Wörterbuch*.

Most importantly, in 1837 the Irish-born English mariner, geodesist and geomagnetist Edward Sabine had collated and rationalized all available geomagnetic intensity observations and published the first global intensity chart. As he had explained in his report to the British Association for the Advancement of Science, the chart incorporated "753 distinct determinations at 670 stations widely distributed over the Earth's surface." But he freely admitted that even this wealth of data was less than ideal, "leaving, it is true, much still to be desired."

Sabine gave a critical evaluation of each of the 753 determinations, which dated back to the explorations and publications of von Humboldt and Hansteen. The data, all given in von Humboldt's relative intensity units, included measurements made in North America by a Scottish botanist, David Douglas (after whom the Douglas fir is named), and many made by Sabine himself on his own numerous voyages. He credited d'Entrecasteaux and de Rossel with having made the very first intensity measurements, but rather than use measurements from the early 1800s he chose instead those made in the 1830s by Captain Robert Fitzroy on his epic five-year voyage of the *Beagle* with Charles Darwin.

Gauss interpolated the data from these charts to estimate values of declination, inclination and intensity at twelve evenly spaced locations around each of seven lines of latitude—a grid of eighty-four locations. He then worked out a series of equations, 168 in all, from which he could calculate the numerical coefficient of each spherical harmonic function in his model. Solving these equations to obtain the twenty-four coefficients was a formidable task; with only pencil and paper it must have taken his team of "calculators" many days to complete.

In the case of the gravitational potential the first term in the spherical harmonic analysis, the one representing a uniform spherical mass, was easily the most important. However, since magnetic monopoles (magnets with only one pole) do not exist, in the geomagnetic potential this term was zero.

By far the biggest contribution in Gauss's analysis came from the second term, which represented a geocentric axial dipole. This

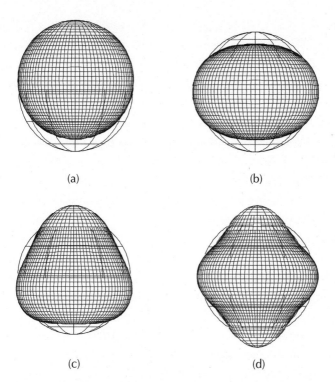

(a)   (b)

(c)   (d)

The principle of spherical harmonic analysis is to describe the potential of the magnetic (or gravitational) field at any particular time as a sum of terms of decreasing size and importance. Shown here are pictorial representations of some of the terms that are symmetric about the rotation axis: (a) the biggest term—the geocentric axial dipole; (b) the second degree, or quadrupole, which is much smaller; (c) the third degree, or octupole, which is smaller again; (d) the even smaller fourth degree term. (Not drawn to scale.)

term was five times bigger than any other, and vindicated Gilbert's original hypothesis based on his observations of his uniformly magnetized terrella.

The next biggest terms were the ones that, when added to this, represented a geocentric but tilted dipole. Taken together, they represented a dipole equivalent in strength to about seventy thousand million billion of the bar magnets typically found in school physics laboratories, with poles at 77°84' N, 296°30' E in northwestern Greenland and 77°8' S, 116°30' E in Antarctica. Since Gauss's time, this part of the field has been known as the dipole field.

All the other terms in Gauss's analysis, the so-called "higher degree" spherical harmonic terms which represented more complicated, "non-dipole" features of the magnetic field, very rapidly became smaller and smaller.

One of the best ways to test a mathematical model is to see how well it can reproduce actual measurements. Hansteen had done this with his four-pole model and now Gauss did the same. Using the twenty-four numerical coefficients of his model, he calculated values of declination, inclination and intensity at the locations of the original observations. The average difference between intensities calculated from Gauss's model and the observed intensity measurements was 0.046 (von Humboldt's relative intensity units), or less than five percent. The average differences in declination and inclination were 1.5° and 1.1°, respectively. Given the small amount of data, the poor distribution of the observation sites around the globe, and the fact that all the calculations had been carried out manually, this was astonishingly accurate.

Gauss's spherical harmonic analysis provided a neat and elegant mathematical way of describing Earth's magnetic field. (It would evolve into the present-day International Geomagnetic Reference Field, in which the numerical coefficients are known as "Gauss coefficients.") However, it brought scientists no closer to understanding what actually caused this magnetism.

Gauss himself seemed curiously uninterested in this. Being at heart a mathematician, he was motivated more by the analytical description offered by the new mathematical methods than by the need for a physical explanation of the phenomenon. He did, however, point out that when better data were available it should be possible to use spherical harmonic analysis to distinguish between internal sources and any possible external ones.

Gauss was sure that the primary source of Earth's magnetism would turn out to be internal, but compared with the elegance of his mathematics his physical reasoning was rather unsophisticated. Curiously, nearly twenty years after Ørsted's discovery that a magnetic field could be the result of an electric current, Gauss stuck steadfastly to the idea of permanent magnetism. However, he did have difficulty accepting the notion of great permanent dipoles deep within the Earth, revolving or precessing fast enough to produce the observed rates of change of declination and inclination—and now intensity—of the magnetic field.

He suggested instead that iron particles in the crust were the source of Earth's magnetic field. In accordance with contemporary notions of a fluid-filled Earth, he reckoned that the solid crust must be relatively thin, but that it was gradually thickening as the fluid beneath solidified on to it, becoming magnetized in the process and so causing local changes in the direction and intensity of the magnetic field. He supported this theory with the observation that the geomagnetic intensity was greatest at the poles where the temperature was lowest, and where you might therefore expect the magnetized crust to be thickest.

Gauss's spherical harmonic analysis of the geomagnetic field was published in 1838 in his *Allgemeine Theorie des Erdmagnetismus*, but by then Gauss was already heavily involved in another project, the establishment of Göttingen Magnetische Verein, a global network of geomagnetic observatories.

By the late 1820s Alexander von Humboldt had become absorbed in studying the rapid geomagnetic time variations that the London instrument-maker George Graham had first noticed in 1722, and that had won Charles-Augustin de Coulomb the prize of the Paris Académie des Sciences in 1777. The disturbed days, which Celsius and Hiorter had correlated with auroras, interested von Humboldt most. He had studied them briefly in 1806, naming them "magnetic storms," but now he made it his mission to carry out a more systematic study of these strange high-frequency disturbances by obtaining simultaneous magnetic measurements from as many different locations as he could.

He began with his own geomagnetic observatory, which he built entirely of non-magnetic materials in the Berlin garden of his friend Abraham Mendelssohn, father of the famous composer, and here he painstakingly collected measurements hour by hour, day and night. He went on to draw all his European colleagues into the project, including François Arago in Paris and Gauss, who co-ordinated a series of simultaneous measurements at observatories in Germany, Sweden and England. He even persuaded Tsar Nicholas I of Russia to build ten new observatories—from Saint Petersburg to Sitka in Alaska (then in Russian hands), and from Arkhangel in the north to Beijing, where the observatory was erected in the grounds of the Russian Orthodox monastery.

Next, he petitioned the president of the Royal Society to set up observatories in British territories the world over, particularly near the equator, at high latitudes, and in the southern hemisphere. This was successful: in 1839 Her Majesty's Government dispatched two ships under the command of James Ross, a naval officer who had located the north magnetic pole—the place where the dip needle came to rest absolutely vertical—on a previous voyage to the Arctic in 1831. On board one of the ships was Edward Sabine, who was charged with overseeing the establishment of magnetic observatories at Toronto, Saint Helena, the Cape of Good Hope and Hobart (then called Hobarton) in Tasmania.

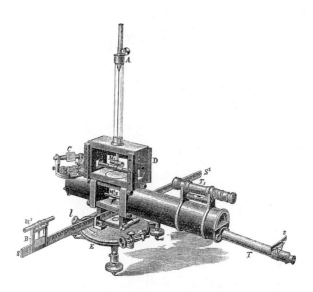

Magnetic observatories relied on accurate instruments to gather data. This magnetometer was used to measure declination and the horizontal component of intensity of Earth's magnetic field, particularly in Britain, the British colonies and the United States from the mid-nineteenth to mid-twentieth centuries. It was designed by Francis Ronalds, director of the Kew Meteorological Observatory in London.

Sabine's background and credentials had made him the natural choice to lead the British front in what would become known as the "Magnetic Crusade." He was now a major of the British Admiralty, and shortly to receive a knighthood. As well as writing a long review of Hansteen's work, and an even longer report to accompany his charts of global geomagnetic intensity, he had carried out land-based magnetic surveys, first in his native Ireland, then in Scotland and England. He had also been involved in several expeditions in search of the elusive Northwest Passage, which was thought to cut through the icy reaches of the Arctic between the Atlantic and Pacific oceans, and had been the astronomer on Ross's expeditions.

The Kew pattern dip circle. Until the mid-twentieth century this instrument was widely used to measure the inclination to the horizontal of Earth's magnetic field at geomagnetic observatories.

When the Board of Longitude was finally disbanded in 1828, having long since fulfilled its goal of achieving accurate determination of longitude at sea, the Admiralty still required a number of scientific advisers, and Sabine, along with Michael Faraday and a physicist called Thomas Young, had been appointed. Sabine had also been a prominent player in the 1831 foundation of the British Association for the Advancement of Science.

The establishment of magnetic observatories proceeded apace. The British East India Company quickly established another four—in Madras, Bombay, Singapore and the Himalayas—and by 1841 Göttingen Magnetische Verein could boast that a total of fifty stations had made observations during one or more of Gauss's selected time intervals. Of these, thirty-five were in Europe, six in Asia, two in Africa, three in North America and four in and around Australia and New Zealand. Curiously, there were no observatories in Central or South America, where von Humboldt had begun his geomagnetic studies.

Early on, Gauss had advocated recording only declination, considering the accuracy of inclination and intensity measurements to be too poor, but by the time Edward Sabine set up the British observatories all three elements were being routinely recorded. In his 1857 report to the Royal Society "On what the Colonial Magnetic Observatories have accomplished," Sabine was able to claim that these measurements represented "not partially, but completely, the whole of the magnetic affections experienced at the surface of the globe." The data would, he said, be useful in studying two aspects of geomagnetism: "the actual distribution of the magnetic influence over the globe, at the present epoch, in its mean or average state;" and "the history of all that is not permanent . . . momentary, daily, monthly or annual change and restoration; or in progressive changes . . . continually accumulating in one direction."

Sabine became immersed in his observatories and the data pouring out from them. In 1841 the Royal Society took over from Göttingen Magnetische Verein the responsibility for collating all the geomagnetic observatory data, and Sabine took on the job of analyzing it. After several years he eventually began to see a pattern in the frequency of magnetic storm activity. He noted that at Toronto, an observatory that consistently produced high-quality results, the disturbances decreased markedly between 1841 and 1843, but from 1843 to 1848 they increased in number again, by a factor of about three.

At the same time, Samuel Heinrich Schwabe, a German pharmacist and amateur astronomer—he had an astronomical observatory on the roof of his house—had been watching the little dark patches that appeared on the face of the sun from time to time. He had, in fact, been observing these sunspots for a great many years while pursuing his real mission of searching for an inner planet, "Vulcan," which he believed would show up as a spot passing across the face of the sun. Schwabe never found Vulcan of course, but he did notice that the number of sunspots followed an eleven-year cycle, with the greatest number occurring in 1828, 1837 and 1848, and the fewest in 1833 and 1842.

No one had yet explained what sunspots were, but to Sabine the correlation with terrestrial magnetic storms was unmistakable: the sun and its cycle of spots seemed to be somehow responsible for von Humboldt's magnetic storms. But how could this be—across 150 million kilometers of empty space?

During the first half of the nineteenth century the silent voice of the magnetic needle had certainly yielded huge amounts of data. Voluminous reports and treatises had been written, and complex mathematical representations constructed. Yet by the end were scientists really any closer to understanding the internal workings of the Earth? Gauss was convinced that the seat of Earth's magnetism was internal. But Sabine's latest results reinforced those of Celsius and Hiorter, and once again seemed to point to an influence from the heavens. Von Humboldt had commented:

> The phenomena of Earth's magnetism, in its three forms of variation, dip and intensity, have of late years been examined with great care, in the most different zones, by the united efforts of many travelers; and there is scarcely any branch of the physical knowledge which, in so small a number of years, so much has been gained towards an acquaintance with its laws, though not perhaps with its causes.

How right he was. The application of science to the study of the Earth's interior was only just beginning. An answer to the puzzle of magnetism would necessitate solving many of the planet's inner secrets and assembling the solutions, one by one, like the pieces of a jigsaw.

The Core of the Matter

Ancient civilizations could only speculate about the nature and shape of the world they lived on . . . Early in the twentieth century it became evident that the interior of the Earth has a . . . structure like that of an onion.

— WILLIAM LOWRIE, 1997

Since 1968, when the Apollo 8 astronauts beamed back the first breathtaking views of Earth, a shimmering globe of blue, green and white, no one can have seriously held on to the notion that our world is flat. Interestingly, though, evidence of the shape of the Earth had been there for all to see since the beginning of civilization, and had certainly been recognized by Aristotle around 350 BC and possibly by Pythagoras 200 years earlier. During a lunar eclipse, as the Earth passes directly between the sun and the moon, a circular arc of shadow is seen to creep across the bright face of the full moon until the lunar disc is completely masked and turns a dull red color. The circular shadow can mean only one thing—the Earth, like the sun and the moon, is spherical.

A hundred years after Aristotle, Eratosthenes, the head librarian at the great library of Alexandria in Egypt, became the first person to estimate the size of the Earth. He employed a clever, sundial-like technique. He knew that at Syene, which lay on the tropic of Cancer near the modern city of Aswan, 5000 stadia due south of Alexandria, the sun was exactly overhead at noon on midsummer's day. By measuring the length of a shadow cast at Alexandria at noon on midsummer's day, Eratosthenes estimated that this distance of 5000 stadia represented one-fiftieth of the circumference of the Earth. If Eratosthenes was using the common Attic stadion, which equals about 185 meters, his result translates to a radius a little over 7000 kilometers. Some scholars think he may have been using the less common Egyptian stadion of 157.5 meters, which would give a radius of 6320 kilometers, or within one percent of the actual value. Either way, the result was remarkable for such a simple experiment.

Later, thanks to Copernicus, Galileo, Kepler and Newton, came the realization that Earth was just one of a number of planets revolving in almost circular orbits around the sun, while at the same time rotating on their own axes. In Newton's time only six planets were recognized—Mercury, Venus, Earth, Mars, Jupiter and Saturn—the same six "wandering stars" that had been known to ancient Greek astronomers. Later, following the development of the astronomical telescope, Uranus and Neptune would be discovered, and in 1931 the American astronomer Clyde Tombaugh would detect Pluto (whose status as a planet was rescinded in 2006).

Compared with what could be seen looking out into space, even by the beginning of the twentieth century little was known of what lay within planet Earth. The ancient Greeks had held that the universe comprised four fundamental elements—earth, water, air and fire—all of which they could see in abundance on the surface of the planet. But although they apparently distinguished different materials on the basis of their physical properties—lodestone and amber, for example—they do not seem to have made much progress in investigating the Earth's underlying constitution.

This was hardly surprising: by 1600 William Gilbert was still commenting on the inaccessibility of Earth's interior to direct sampling and experimentation. Even today, scientists have drilled a mere twelve kilometers into the outermost layers of the planet; the rest remains the realm of science-fiction writers and remote-sensing methods of investigation.

By the beginning of the twentieth century, however, geophysicists had begun to realize that gravity, geomagnetism and earthquakes all yielded valuable clues to the make-up of the planet if only these clues could be deciphered and interpreted. In the eighteenth century, Edmond Halley had envisaged Earth's interior as a series of shells that nested inside one another like Russian matryoshka dolls and were scaled in harmony with the heavens. Although this model never really caught on, the basic idea of a concentric layered structure, comprising an inner nucleus or core and a rigid outer shell or crust, would be the starting point for almost all later models. However, intense debate was to develop about the composition and physical state of the various layers.

The first indication that Earth might not be a homogeneous lump of rock had appeared in 1798, when Henry Cavendish, the reclusive professor of physics at Cambridge University, announced that "the density of the Earth comes out 5.48 times greater than that of water." This meant that, on average, one cubic meter of Earth had a mass of 5480 kilograms. (Cavendish had actually made an arithmetical mistake: his figure should have been 5450.) By the same calculation, the mass of the whole Earth was about six million million million million (6×10^{24}) kilograms.

Cavendish's result—that Earth's density was five and a half times that of water—came as a huge surprise since the densities of rocks found at the Earth's surface averaged only about two and a half times that of water, and were rarely more than three times the density.

Had Cavendish made a mistake? It seemed unlikely. Thirty years earlier the Astronomer Royal, Nevil Maskelyne, had also tried to estimate Earth's density using a quite different method. He had measured the deflection of a plumb line towards a cone-shaped Scottish mountain, Schiehallion, and obtained a density value four and a half times that of water. Although Maskelyne's and Cavendish's values were different, both were high enough to indicate there was more to the Earth than just its surface rocks.

Although often dubbed his attempt to "weigh the world," Cavendish's experiment was actually designed to test Newton's law of gravitation, then more than a century old, and to determine the still unknown universal gravitational constant, G. To do this he had set out to compare the force of gravity between two massive lead spheres and the gravitational force that the Earth exerted on each of them—in other words, their weight.

He had inherited the design for this experiment from Reverend John Michell, a minister in the Church of England and amateur scientist, who had died before being able to complete the work himself. It comprised a huge torsion balance similar to those that Cavendish and Coulomb had used to investigate the electrostatic and magnetic forces, but much bigger: the main beam was nearly two meters long. At each end of the beam was fixed a two-inch diameter, 1.6-pound lead sphere, and about nine inches from each of these Cavendish suspended a twelve-inch, 348-pound sphere. He then meticulously measured the tiny forces acting between each pair of spheres. To minimize extraneous disturbances, the whole apparatus was sealed in a room of Cavendish's house, with Cavendish operating from outside and observing through a tiny window.

Henry Cavendish, the Cambridge University professor of physics whose experiment to "weigh the world" led to the conclusion that the Earth has a heavy, iron-rich core.

Before long the Cavendish experiment had become the standard way to measure gravitational forces in the laboratory, and numerous repetitions revealed no error in Cavendish's original result. It was clear there was some extremely dense material within the Earth.

Around this time scientists began to question the physical and chemical properties of the rocks found on the Earth's surface, and the processes by which these rocks had been formed, deformed and reformed through time. This would be the starting point of a new scientific discipline: geology. To explain the fluid magma spewed out from volcanoes, earthquakes and other geothermal activity, some geologists argued that Earth must have a molten interior, and only a thin solid crust. This was consistent with the observation—for example, in deep mine shafts—that temperature increased with depth, leading to the conclusion that the melting points of common rock-forming materials might be reached just a few tens of kilometers down.

Many physicists opposed this theory. Ampère, for example, argued that during the course of its orbit around the Earth the moon would raise tides in the magma, as it did in the waters of the oceans, and that these tides would tear apart such a thin crust. Poisson disputed whether such a crust could form over a predominantly molten interior in the first place. He believed that very early in Earth's history, when the whole planet was still molten, cooler blobs of fluid would have sunk because of their higher density, so any solidification would have taken place from the inside out, rather than the outside in.

Two schools of thought emerged. The first favored a largely molten Earth with a thin rigid crust. The second supported a model more like Halley's, in which Earth had a substantial, solid nucleus and a relatively thin fluid layer underneath a rigid crust.

One of the first scientists to address the problem objectively was William Hopkins, a Cambridge University mathematician. Academically, Hopkins was something of a late starter. Having tried his hand at farming and not enjoyed it, he was already twenty-nine and married for the second time when in 1821 he entered Peterhouse at Cambridge to study mathematics. Age proved no barrier: he excelled and graduated seventh "Wrangler"—seventh of the first-class honors students in the year's Mathematics Tripos.

Being married, Hopkins now found himself ineligible for a college fellowship. However, he gained a lectureship in mathematics and also became the university's Esquire Bedell, a largely ceremonial position, the responsibilities of which included carrying the university mace when accompanying the vice-chancellor on official occasions, and carving the roasted joint for him at banquets. Hopkins tutored many of Cambridge's top mathematics scholars, including William Thomson (later Lord Kelvin), Gabriel Stokes and James Clerk Maxwell, and soon became known as "the wrangler-maker."

Through his friendship with Adam Sedgwick, a Fellow of Trinity College and inaugural professor of geology, Hopkins also developed a passion for geology, in particular the application of mathematics and physics to investigations of the Earth's interior. Although often working on shaky assumptions (the reason his work is virtually ignored today), he would lay the groundwork for much of modern geophysics.

It was clear that in the Earth's interior there had to be layers of fluid whose temperature exceeded the melting point of rock-forming materials. Hopkins pointed out that with increasing depth a trade-off would happen: beyond a certain depth the increase in temperature would, other factors being equal, lead to melting—but the increase of pressure with depth would also raise the melting point, and if the pressure were high enough this might result in solidification. In an attempt to determine a "melting-point curve" for the Earth, he enlisted the help of his former student William Thomson, and another physicist, James Joule, who was by then famous for his experiments on heat and energy, to try and measure the melting points of rocks under high pressure. Unfortunately, these experiments proved rather too ambitious: the men reached no firm conclusions and Hopkins turned to other means of detecting fluidity within the Earth.

It had long been known that Earth's rotation axis shifted slightly but regularly in response to the gravitational pull of the sun and the moon. Astronomers had observed two separate effects, which they called precession and nutation. Hopkins figured that details of these tiny angular motions would differ depending on whether the Earth were rotating as a solid whole, or internal fluids were slipping and lagging behind the solid parts. His first set of mathematical calculations—nearly forty pages—was presented in the first part of his "Researches in Physical Geology" in the 1839 *Philosophical Transactions of the Royal Society*, but the calculations had proved inconclusive. Hopkins was still unable to distinguish between the models over which the geologists and physicists were arguing.

However in 1842, in the third part of his "Researches," Hopkins finally came down in favor of a predominantly solid planet, in which volcanic magma became truly fluid only with the release of pressure as it rose through near-surface fissures and cavities immediately prior to eruption. This must have influenced William Thomson, who subsequently declared the Earth to be "as rigid as steel," and made his famously erroneous estimate of the age of the Earth as a few tens of millions of years on the basis that the planet could lose heat only by solid-state conduction.

Many geologists and biologists vehemently disagreed, arguing that it must have taken far longer for the vast thicknesses of sedimentary rocks present on the Earth's surface to have accumulated, or for life to have reached its present degree of evolution.

New information on the shape and structure of the Earth would next come from detailed measurements of gravity at the surface, a study known as geodesy. Even before Isaac Newton's publication of *Principia* in 1687, it had been noticed that a pendulum swung more slowly at the equator than at higher latitudes. In 1672 a French astronomer, Jean Richer, had noted that his pendulum-clock lost two-and-a-half minutes per day at the equator, compared with measurements made before he had left Paris. This meant that at the equator the force of gravity was slightly weaker and "g," the acceleration due to gravity, slightly less than at Paris.

Newton had explained this by suggesting that the planet bulged at the equator, so Richer's pendulum-clock was further from the center of the Earth and therefore experienced a smaller gravitational force, causing it to swing more slowly. The equatorial bulge, Newton had argued, was a direct effect of Earth's rotation. To make something revolve in a circle it must be pulled inwards (centripetal force): the smaller the circle or the faster the rotation, the bigger the pull that is needed. In order to provide this pull, Newton supposed that the Earth became stretched outwards at the equator.

In the eighteenth century, the Swiss mathematician Leonhard Euler had predicted that, as a result of the equatorial bulge, Earth's rotation axis should be displaced a few meters from the axis of symmetry, and should move around it. Like the precession and nutation studied by Hopkins, the detailed dynamics of Euler's perturbation would depend on how mass was distributed within the Earth and how rigid it was—"as rigid as steel," as Kelvin had proposed, or fluid, as many geologists inferred, or somewhere in between. If the Earth were rigid, calculations showed the period of Euler's axial perturbation would be 305 days. However, when the effect was finally observed in 1891 by an American astronomer, Seth Carl Chandler, it was found to be 435 days, or about 14 months.

The effect, now known as Chandler Wobble, meant that latitudes seemed to fluctuate very slightly in a fourteen-month cycle. This showed the planet could not be truly rigid: it must deform very slightly in response to the forces imposed on it. It was, in other words, "elastic."

Just as important as the physical state of Earth's inner regions was the chemical composition of the materials that made it up. Surface rocks were one thing but what could there be deep down, either solid or liquid, that was so much denser than the rocks on the surface?

In 1896 Emil Wiechert, a young geophysicist from Königsberg in Prussia, put forward a simple two-part model of the Earth that he claimed was compatible with all available astronomical and geodetic information. Observations of the precession and nutation of the axis of rotation had by now shown that the densest material must be concentrated very deep down, close to the center of the Earth. To account for this, Wiechert boldly proposed a core of completely different composition from the overlying rocky shell. Since the only materials known to have densities higher than Earth's average value of 5500 kilograms per cubic meter were metals, he suggested that the core must be iron, the most abundant of the stable metallic elements. This made sense. Iron was known to be a common constituent of meteorites, and these were thought to be fragments of planetary bodies that had formed and broken up early in the history of the solar system.

According to Wiechert's calculations, the core had to have a density of 8200 kilograms per cubic meter and occupy nearly fifty percent of the volume of the planet, giving it a radius of some 5000 kilometers. The overlying shell had to have an average density of 3200 kilograms per cubic meter.

Wiechert called this outer layer "der Mantel." Some scientists objected on the grounds that a mantle was a loose, floppy cloak and so this was a poor description, but the word stuck. Meanwhile, Wiechert's calculations would be repeatedly revised and improved. Today we know that the core's radius is about half the total radius of Earth, and so the core occupies only about fifteen percent of Earth's volume, with the mantle making up virtually all the remaining eighty-five percent.

Amid the increasingly complex wobbles and perturbations of geodesy and the ever more detailed measurements of gravity, a new method of remotely sensing Earth's interior had unexpectedly appeared on the scene. Ernst von Rebeur-Paschwitz, a German geodesist, had set up a delicate pendulum at Potsdam near Berlin, and another at the coastal town of Wilhelmshaven, 350 kilometers to the northwest. These pendulums were designed to measure the horizontal motion of the ground due to supposed lunar tides. According to Rebeur-Paschwitz, they would record this information automatically "by the same photographic method as that employed for magnetic observations."

On April 17, 1889, at very nearly but not quite the same time, both instruments were violently shaken by a sudden series of vibrations. Rebeur-Paschwitz was completely mystified until two months later he read in the June 13 issue of the scientific journal *Nature* of a great earthquake that had rocked Tokyo. He would later write:

> Reading the report on this earthquake, I was struck by its coincidence in time with a very singular perturbation registered by two delicate horizontal pendulums at the observatories of Potsdam and Wilhelmshaven.

Taking local time differences into account, he calculated that the earthquake disturbances had taken an average of 3858 seconds—just over an hour—to travel from Tokyo to Germany. As this was an average surface distance of 8264 kilometers, he estimated their speed to have been just over two kilometers per second. Considering this a reasonable figure for waves traveling close to the surface of the Earth, he concluded:

> . . . the disturbances noticed in Germany were really due to the volcanic [*sic*] action which caused the earthquake of Tokio.

Interestingly, this was not a new idea: it had been predicted at least half a century earlier that vibrational waves should propagate not only along Earth's surface but also through its interior, and that their speed might shed light on the physical properties and composition of materials inside the Earth. French mathematicians Augustin-Louis Cauchy and Siméon-Denis Poisson had shown that energy should propagate through flexible solid materials in the form of waves, and William Hopkins had adapted the theory to the case of the Earth, and given a detailed exposition on the "vibratory motions of the Earth's crust produced by subterranean forces" and the "observations required for the determination of the center of earthquake vibrations, and on the requisites of the instruments to be employed" to the 1847 meeting of the British Association for the Advancement of Science in Oxford.

The physics behind this was relatively simple. When an elastic material is compressed, it responds by trying to expand back out. Conversely, when it is stretched (or "rarefied") it tries to contract. If a material is alternately compressed and stretched, it responds by radiating out into the material around it a series of compressions and rarefactions: this motion is called a wave.

Hopkins had predicted that two different types of elastic waves would travel through the body of the Earth. The first, which he called "normal vibrations," would involve alternate compressions and rarefactions, with parts of the material moving forwards or backwards, parallel, or in the opposite direction to that in which the waves were traveling. Today these are known as P ("primary" or "pressure") waves. They are identical to sound waves.

Secondly, Hopkins said, there should also be "tangential vibrations" in which the material would undergo a shearing process, so parts of it moved at right angles to the direction in which the wave was traveling—rather like a wave on the surface of water. Today these are called S ("secondary" or "shear") waves.

Hopkins explained that in the same material a normal (primary) wave would always travel faster than a tangential (secondary) wave, since the speed depended on the compressibility of the material as well as its rigidity, or resistance to shearing motion. If an earthquake took place and the speeds in the Earth of the primary and secondary waves were known, it would be possible to work out, from the time between the waves' arrivals at a certain spot, the distance to the quake's epicenter. He envisaged doing this across a network of observation sites so earthquake centers could be accurately located by triangulation; he even described the requirements of the instruments that would be needed to measure the vibrations.

Furthermore, whereas both types of waves could travel through solid materials, Hopkins pointed out that only primary waves could travel through fluids. A fluid, because it could flow, had no resistance to a shearing deformation and so could not sustain the sort of movement involved in the passage of secondary waves. Put simply, solids could transmit both P and S waves, with the P waves being faster, while fluids could sustain only P waves.

Hopkins went on to predict that the study of earthquake disturbances would eventually enable scientists to reconstruct the paths of seismic waves through the different layers of Earth, and so decipher the planet's hidden structure. Now, at last, Rebeur-Paschwitz had, albeit unintentionally, detected seismic waves, even if they were surface waves, rather than the body waves Hopkins had predicted. He had shown that with suitable instrumentation the waves' travel times could be measured, and from this their speeds could be calculated. This information could then be interpreted to uncover the physical properties of the rocks through which the waves had traveled.

Tragically, Rebeur-Paschwitz contracted tuberculosis and died in 1895 at the age of only thirty-four, before he had a chance to further develop his interest in seismology and the structure of the Earth. However, other scientists had by then recognized the need for a global network of earthquake observatories similar in concept to Gauss's Göttingen Magnetische Verein. Seismology was up and running. Over the next few decades, observations, interpretations and new ideas about Earth's interior would accumulate rapidly.

In 1897, after a particularly violent earthquake in Assam, India, an English seismologist, Richard Dixon Oldham, finally succeeded in separating the arrival signals of P and S waves that had traveled through the interior of the Earth from those of the slower waves that had traveled along the surface. He suggested that the waves' paths and speeds were consistent with the Earth having a glassy or stony mantle, and an iron core that extended from its center to just over half of its radius. Although this was remarkably close to modern estimates of the size of the core, Oldham made some assumptions that would come to be considered dubious and so today his calculation is rarely acknowledged.

It would be Beno Gutenberg, Wiechert's twenty-three-year-old student from Göttingen, who would come to be credited with the discovery of a sudden drop in seismic wave speed about halfway to the center of the Earth. The clue that led Gutenberg to propose such a "low-velocity" core was what has been called the "shadow zone." Assuming that P and S waves radiated in all directions from the source of an earthquake, Gutenberg reasoned that waves traveling along shallow paths would reappear at the surface relatively close to the epicenter, while those traveling deeper down would reappear further away. However, the waves could reach only a certain distance, about 12,000 kilometers from the epicenter, without having to pass into the core.

Gutenberg found waves that entered the core reappeared at Earth's surface much further from the earthquake source than they would have done had they traveled through a completely uniform Earth. In fact, no P (or S) wave arrivals were detected anywhere between about 12,000 and 16,000 kilometers (105° and 143°) from an earthquake's location. He inferred from this that when the waves entered the core they slowed, making their paths steepen, and when they re-emerged their paths were such that they could not reach the surface anywhere within the so-called "shadow zone." From these observations, he calculated that the boundary between the mantle and the core must lie at a depth of about 2900 kilometers and that, on entering the core, seismic waves must suddenly slow to about 65 percent of their speed in the mantle.

Gutenberg supposed both the mantle and the core to be solid. He envisaged a change in composition from a rocklike material with a P wave speed of about 12 kilometers per second to a possibly iron-rich core with a P wave speed of about ten kilometers per second. In this respect, his seismic model of the Earth was in agreement with Wiechert's model, which had been based on density and on geodetic and astronomical observations. Each envisaged two parts, but the boundaries between the two parts lay at quite different depths in the two models.

The question was, were the two scientists looking at the same transition, or were there two? By this time, other geophysicists had revived the idea of fluids deep inside the Earth, and were arguing that the transition (or transitions) corresponded to a change not from one solid state to another, but from a solid mantle to a liquid core. One strong argument for the core being liquid was the growing

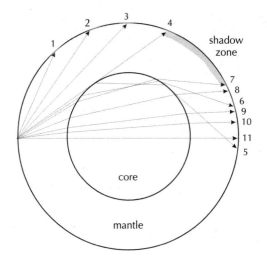

The formation of a seismic shadow zone. Waves entering the core are slowed down and refracted so they reappear at the surface further from the source than they would have otherwise, leaving a shadow zone in which no waves are detected. In the Earth the shadow zone is observed between 105 and 143 degrees from the source of an earthquake.

evidence that *S* waves did not seem to travel through it at all: unlike *P* waves, they did not reappear beyond the shadow zone. An even more compelling argument was that the average rigidity of the Earth, as deduced from tidal measurements of the sort studied by Rebeur-Paschwitz and from the Chandler Wobble, could simply not be reconciled with a completely solid Earth.

It would be 1926 before a pronouncement by a noted British geophysicist would finally settle the argument. Harold Jeffreys, the author of the famous and much reprinted textbook *The Earth*, had improved Wiechert's calculations and eventually come to the conclusion that Wiechert's discontinuity in density and Gutenberg's seismic discontinuity were one and the same. Furthermore, the discontinuity corresponded to changes both in composition (from

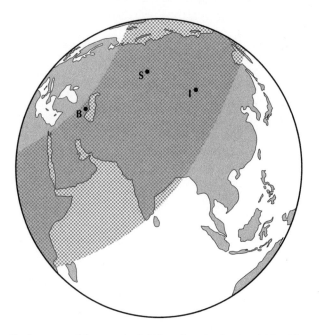

Shadow zone of the magnitude 7.8 earthquake that struck Murchison, New Zealand on June 17, 1929. New Zealand lies out of sight on the bottom right reverse side of the globe. Cast by Earth's core, the shadow zone, within which no *P* or *S* waves were expected, covers the area between 105° and 143° from the epicenter. Baku (B), Sverdlovsk (S) and Irkutsk (I), all lay within it.

rocky to iron-rich material) and in phase (from solid to liquid). He therefore declared that Earth's core was "truly fluid."

It seemed the Earth's interior was at last understood. Geodesists, astronomers, seismologists, geologists and evolutionists were all in agreement. Even physicists, having discovered in radioactivity both another heat source within the Earth and a new means of dating rocks, revised their estimate of the age of the Earth to several thousand million years and joined the party.

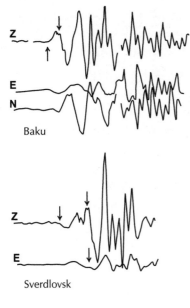

Baku

Sverdlovsk

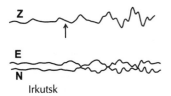

Irkutsk

Seismograms from Baku, Sverdlovsk and Irkutsk all show unexpected *P* wave arrivals (indicated by arrows) from the New Zealand earthquake of June 17, 1929, a finding which led to the discovery of Earth's inner core. The weak *P* waves are followed by much bigger surface waves.

The picture was not quite complete, however. The seismologists had one more surprise up their sleeves.

Inge Lehmann was an exceptional woman. Born in 1888, she came from an unusually liberal background that prepared her to hold her own in the male-dominated scientific world of the early twentieth century. As a child in Denmark, Lehmann had attended a remarkably progressive coeducational school run by Hanna Adler, an aunt of the famous physicist Niels Bohr. Here all pupils, boys and girls, were taught together and studied the same subjects: reading, writing, arithmetic, football, cookery and needlework. There was no segregation or discrimination.

In 1925, several years after graduating from the University of Copenhagen with a master's degree in mathematics, Lehmann had begun work on earthquakes and established a network of seismological stations in Denmark and Greenland. By 1928 she was chief seismologist at the Royal Danish Geodetic Institute and had access to data from stations all over northern Europe.

In 1929 a magnitude 7.8 earthquake hit Murchison, a tiny settlement in the South Island of New Zealand. Lehmann's network of observatories covered a large part of the shadow zone, where no seismic waves were expected. Nevertheless, Lehmann detected faint P wave arrivals from Murchison. How could this be? There was no going back on the conclusion that the core was liquid—more and more evidence was accumulating from disciplines other than seismology. The only possible answer was that there was something inside the liquid core—something with a higher seismic wave speed—that had reflected back the P waves so they eventually emerged within the shadow zone.

Matching Weichert's boldness, Lehmann now proposed that Earth had an additional inner core. She went further. Assuming uniform P wave speeds of ten kilometers per second in the mantle and eight kilometers in the outer core, and adjusting the radius and the P wave speed in her inner core to fit the observed arrival times, she estimated that the inner core had a radius of 1400 kilometers and a P wave speed of ten kilometers per second. She named these core waves P' (P-prime), and her original paper on their discovery is entitled simply "P'."

The discovery that not only did Earth have an iron-rich core, but the outer part of this core was liquid, and so both mobile and

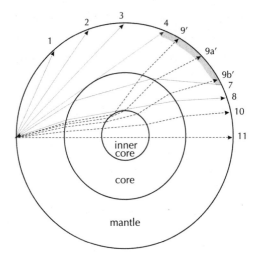

Above: Seismic ray paths through the Earth. The detection of weak *P* wave arrivals in the shadow zone led Lehmann to conclude there is an inner core in which the seismic waves travel faster so they are eventually refracted into the shadow zone.

Below: The main features of Earth's internal structure: a solid iron-rich inner core, a liquid iron-rich outer core, a rocky mantle, and a rigid but brittle crust.

electrically conductive, opened up a new era of geomagnetism. Suddenly, permanent magnetism was no longer needed to explain Earth's magnetic field: electric currents flowing in the metallic core might be generating the field. But if this were the case, why would there be electric currents in the core in the first place?

Faraday's electromagnetic induction provided a clue. If, for some reason, the conducting core fluid moved through a magnetic field, electric currents would be induced in it, in exactly the same way as Faraday had induced currents in his early experiments. These electric currents would, in turn, produce magnetic fields. In the right circumstances, these secondary, induced fields might reinforce the original field. In other words, as long as energy was available to drive the original fluid motion the process could be self-sustaining.

To Gilbert and Halley, even to Faraday and Maxwell, the idea of such a "magnetohydrodynamic" process operating inside the Earth would have been totally unimaginable. By the early twentieth century it was just barely credible. However, by the time the dynamo theorists got seriously down to work a few decades later, several amazing new discoveries about the history of Earth's magnetic field were to make their task even harder.

Reading the Rocks

Intuitively . . . one regards the Earth's magnetic field as
an inherent property of such grandeur that reversals of
its sense are difficult to grant.

— JOHN GRAHAM, 1952

Scientists continually seek new and better observations so they can
create ever more detailed and accurate explanations of the natural
phenomena they study. This is never truer than of geomagnetists.
Over time, as more observations accumulated of Earth's magnetic
field, a pattern of increasingly complex variations was added to
Gilbert's original concept of a geocentric axial dipole.

By 1900 it had become apparent that secular variation—the
gradual changes in direction and intensity of Earth's magnetic field
occurring year by year, decade by decade—was a persistent feature.
The 300 years of declination and inclination measurements collected at London and Paris seemed to trace out two-thirds of a loop.

It was tempting to speculate that the loop would complete itself in another 150 years, and perhaps even repeat itself again in the next 400 to 500 years. Was there a regular cycle to this?

Scientists are also notoriously impatient, and 150 years was far too long to wait for an answer. What was needed was a prehistoric record so they could find out what had happened in the past.

It was Joseph Fournet, the inaugural professor of geology at the University of Lyons in France, who first suggested that magnetized rocks might hold the answer, but the beginning of paleomagnetism is usually credited to his contemporary Achille Delesse, who held the chair of mineralogy at nearby Besançon. In 1849 Delesse reported that certain volcanic lava flows were magnetized—not strongly and randomly like von Humboldt's lightning-stricken rocks, but regularly, if more weakly, in the direction of Earth's magnetic field. Even when removed from their natural position and orientation, samples of the lavas retained a stable magnetization.

A few years later an Italian physicist called Macedonio Melloni showed that lava flows from Mount Vesuvius and the Campi Flegrei (Phlegraean Fields) area west of Naples were also magnetized parallel to Earth's magnetic field. Melloni took samples of the lavas back to his laboratory, heated them until they glowed red and then cooled them in the laboratory's magnetic field. He found they lost their original magnetization on heating, but became remagnetized parallel to the laboratory field when cooled again. This confirmed his supposition that the lavas' original magnetization had been acquired during their initial cooling in Earth's magnetic field following the eruption of Mount Vesuvius.

This discovery of fossil magnetic records preserved in lava flows seems to have then gone largely unnoticed for nearly fifty years. In the meantime another Italian, Giuseppe Folgheraiter, began to wonder whether archaeological artifacts such as pottery and bricks, which had been fired by humans rather than nature and had cooled in the local magnetic field, might carry a stable magnetization. If so, they could be just what was needed to extend the record of secular variation backwards in time. One feature that attracted Folgerhaiter to such artifacts was that they could often be dated quite accurately from their style and decoration, and sometimes from documentary records associated with them.

Folgheraiter was not the first to recognize the magnetization of bricks and clay objects. As early as 1690 Robert Boyle, a contemporary of Newton and Halley, had found that a brick heated in a fire and then cooled became magnetized in the direction of the surrounding magnetic field, while in the 1860s another Italian, Gheradi, had made a systematic study of Egyptian and Italian pots and demonstrated the stability of their magnetization. Gheradi had also shown that the magnetization of certain building bricks could affect sensitive magnetic measurements, and warned against the use of such bricks in the construction of geomagnetic observatories.

Folgheraiter's work, published in 1894, involved sampling figured red-clay Greek vases and urns that dated back to the sixth century BC. Folgheraiter reasoned that if the orientation of a vase in a firing kiln were known, the magnetization it had acquired on cooling should give a record of the geomagnetic field at the time of firing. He sampled and examined vases of different ages, and drew up an inclination record for the period 800 BC to 100 AD.

Intriguingly, the earliest of Folgheraiter's results showed a negative inclination, suggesting an anomalous or curious field direction at the time of the vases' magnetization. Scientists still disagree as to whether Folgheraiter mistook the orientation in which the pots were fired and so got an invalid result, or whether the geomagnetic field really was anomalous at the time and location of their firing.

Whatever the truth, these pioneering archaeomagnetic and paleomagnetic studies set the scene for some startling discoveries in the twentieth century. Almost immediately, Folgheraiter's success on man-made clay artifacts inspired Bernard Brunhes, a brilliant young French physics professor, to search out clays that had been baked by natural processes. In the early 1900s he and his assistant, Pierre David, went to the Massif Central region of France looking for layers of clay that lay directly under ancient lava flows. They reckoned that not only would the clay have been baked by the hot lava flow during the eruption, but both clay and lava should have then cooled together and so carry the same record of the geomagnetic field direction at the time.

Brunhes and David sampled a number of lava flows and their underlying baked clays. For the most part the results met their expectations. However, one result really caught their attention. It came from a site at Pontfarein, near the town of Saint-Flour, where road-building authorities had made a hundred-meter-long road cut. Brunhes and David had collected several samples of the baked clay from there, but only two of the basalt lava flow as it had been particularly hard and difficult to cut. One of these lava samples turned out to have been struck by lightning and so was useless, but the magnetic direction of the other agreed very well with the average result from the baked clay samples.

However, there was something very strange. The declinations Brunhes and David obtained from the clays and lava were 148° and 154° respectively, while the inclinations were –74° and –76°. The magnetizations of the rocks pointed approximately to the southeast and angled upwards, while the geomagnetic field at Pontfarein, as at most locations in the northern hemisphere, was roughly northwards and inclined down. In other words, the rocks were magnetized in almost the opposite direction to the local magnetic field.

Brunhes and David were astounded. They could hardly believe their results, but the consistency was too good to occur by chance. In 1906 Brunhes nervously announced to the world:

Bernard Brunhes, who, with his assistant Pierre David, discovered rocks at Pontfarein that were magnetized in the opposite direction to the local magnetic field. They deduced that central France had once been closer to the south magnetic pole than to the north—a concept that was to confound the scientific world for the next fifty years.

. . . *qu'en un moment de l'époque Miocène aux environs de Saint-Flour, le pôle Nord était dirigé vers le haut: c'est le pôle Sud de la Terre qui était le plus voisin de la France centrale.*

(. . . at a moment in the Miocene epoch, in the region of Saint-Flour, the north pole was directed upwards; it is the Earth's south pole that was closest to central France.)

The obvious interpretation was that at some "moment" about six million years ago Earth's magnetic field had been upside-down. The idea was so surprising that fifty years later scientists would still be struggling to understand if such a thing could happen.

Curiously, neither Brunhes nor David published any further on this subject. As well as his job as professor of physics at the nearby University of Clermont-Ferrand, Brunhes' time was taken up rebuilding the observatory at Puy-de-Dôme, where he had been director since 1900. He expanded the observatory considerably, and extended its original meteorological portfolio to include both geomagnetic and seismological observations. One night in 1910 the local gendarmerie came across a man lying unconscious on a street in Clermont. Thinking he was the victim of an assault but finding no obvious injuries on his body, they searched him, and on discovering his identity carried him home. Bernard Brunhes died the next day, probably from a brain hemorrhage. He was just forty-three.

Almost a hundred years after Brunhes' discovery of reversely magnetized rocks, three French paleomagnetists—Carlo Laj (supposedly a distant relative of Brunhes), Catherine Kissel and Hervé Guillou—would relocate and resample the Pontfarein section. Using modern laboratory techniques, they would obtain magnetization directions for both the lava flow and the baked clay that were indistinguishable from Brunhes' results. In addition, they would date the lava at six million years.

Despite Brunhes' stunning results, it would take a long time for the geophysical community to be convinced that the polarity of the Earth's magnetic field had reversed. French physicist Paul Mercanton was the first to continue Brunhes' work. Mercanton reasoned that if the main, dipole part of the geomagnetic field really had reversed its polarity, reversely magnetized rocks should be found all round the world, and he set out to look for them. Beginning in 1910, he published a series of papers in which he described both "normal" and "reversed" directions of magnetization from lava flows estimated to be up to sixty-five million years old in Spitsbergen,

Greenland, Iceland and the Scottish islands. Mercanton cautiously noted that his results supported the theory of geomagnetic field reversals, and urged other scientists to continue the search for reversely magnetized rocks, particularly in the southern hemisphere. He went to the extent of acquiring some lava samples from Australia, but the results were disappointingly inconclusive. Nevertheless he ended his 1926 paper with an idea that would prove to be both inspired and inspiring:

> . . . if a link exists between the rotation axis of our globe and its magnetic axis, the considerable displacements of the magnetic axis which our researches would discover would unexpectedly corroborate the large displacements of the axis of rotation argued for by A. Wegener.

From this time on, the controversy over geomagnetic reversals would become entwined with another equally contentious theory. As early as 1596 a Dutch mapmaker named Abraham Ortelius had observed that there was a remarkable fit between the coastlines of Africa and South America. Europe and Africa, he suggested, had been "torn away" from America by great earthquakes and floods.

The idea resurfaced in the early nineteenth century, courtesy of Alexander von Humboldt among others, but it was not until 1912 that a German meteorologist, Alfred Wegener, formally put forward the theory of what he called "continental drift." In addition to the way the coastlines matched, Wegener pointed out that various fossil species were to be found on both sides of the oceans, as were similar rock formations and geological structures. There was also evidence of huge-scale climate change: fossils of tropical plants had been found in Antarctica and evidence of glaciation could be seen in South Africa.

Wegener extrapolated back to a time around 200 million years ago when, he proposed, all the continents had been joined together in one mega-continent. He named this Pangaea from *pangaia*, a Greek word meaning "all Earth." In 1937 a South African geologist, Alexander Du Toit, further refined the thesis by suggesting that Pangaea had first split in two, forming Gondwana in the south and Laurasia in the north, and that after drifting apart these two supercontinents had further broken up into the modern continents.

These twin theories—geomagnetic reversals and continental drift—now sat simmering side by side on the back-burners of science for several decades. Each slowly gathered support but also suffered setbacks. The problem was that no one could come up with viable explanations as to why either phenomenon should occur. And given that both theories were wildly at odds with conventional wisdom, it was hard for many people to take them seriously at all.

Mercanton's work was eventually noticed and followed up by a Japanese paleomagnetist. Motonori Matuyama studied more than one hundred lava flows from Japan, Korea and Manchuria, and made a significant finding: all his youngest lava flows were normally magnetized, while his reversely magnetized samples all came from underlying, older flows.

Two streams of paleomagnetic research continued through the Depression years of the 1930s but neither made much progress on the mystifying polarity reversals. Instead they aimed at getting a better picture of secular variation. In France, Émile and Odette Thellier concentrated on fired archaeological artifacts, trying to extract information on the intensity of the paleomagnetic field, as well as its direction. By showing that the strength of the magnetization acquired by an artifact as it cooled after firing depended directly on the strength of the local magnetic field, they were able to estimate changes in the strength of the ancient magnetic field.

Meanwhile, in the United States the pace was also picking up. The Carnegie Institution's Department of Terrestrial Magnetism in Washington, DC was already well known for its study of historical observations of the magnetic field. Now a paleomagnetism group emerged, and members busied themselves investigating whether sediments that had gradually accumulated on the seafloor might carry a continuous record of changes in the direction of the geomagnetic field—something like a natural tape recording.

In 1940 Sydney Chapman of Imperial College, London and Julius Bartels of the Potsdam Geophysical Institute published a landmark two-volume textbook, *Geomagnetism*. Curiously, though, paleomagnetism rated scarcely a page: the focus was almost entirely on direct observations of Earth's magnetic field and their interpretation, and there was no mention of the work of Brunhes, Mercanton or Matuyama. However, Chapman and Bartels did describe in detail the case of the Pilanesberg dyke system in the Transvaal region of South Africa.

Dykes are intrusions of lava that have squeezed into fissures or faults in older rocks. Fourteen or more almost vertical dykes, each about 160 kilometers long and reckoned to be a hundred million years old, radiated from a point about 120 kilometers northwest of Pretoria. When a magnetic field survey had been conducted over each dyke, a strong positive anomaly had been recorded. A positive anomaly could only mean that the dykes were permanently magnetized in the opposite direction to the geomagnetic field.

In the first volume of *Geomagnetism*, Chapman and Bartels appeared to accept this interpretation, but in the second they seemed to hedge their bets, writing:

> Certain geological evidence that suggests a possible complete reversal of the Earth's field during past ages is usually disregarded as unreliable—a view which is perhaps too dogmatic.

Chapman and Bartels' book would be one of the last Anglo-German collaborations for some time. The outbreak of World War II would affect scientific research in Britain and much of Europe. Non-strategic research ground to a virtual standstill, while technology, including the development of materials and instruments that might have a military application, accelerated beyond all imagination. Ironically, this would have positive spin-offs for geomagnetism: the production of rugged electronic devices, and of materials that are easily magnetized and extremely stable, would have an enormous impact in the post-war years.

In France, a keen young physicist named Louis Néel now took the first steps towards an explanation of how and why rocks and baked clays retain remarkably strong and stable magnetizations over hundreds of millions of years of geological time. Néel's studies had begun in 1928 at the University of Strasbourg's Physics Institute, which had been established by Pierre-Ernest Weiss to focus on magnetic research. Weiss and his fellow physicists Pierre Curie and Paul Langevin had already begun to rewrite the physical theory of magnetism. Their pioneering work had dovetailed with revolutionary late nineteenth- and early twentieth-century discoveries in atomic and quantum physics. It would form the basis of Néel's theory of rock magnetism.

Materials that can hold on to a stable, long-term or "remanent" magnetization are generally called ferromagnets. Iron and lodestone (magnetite) are common examples, but other natural minerals such as the rusty red iron oxide hematite and some iron sulfides are also ferromagnetic.

Faraday had suggested that most other materials should be classified as "paramagnetic" or "diamagnetic" because they acquired weak "induced" magnetization, either parallel or opposite to an ambient magnetic field, and lost it as soon as they were removed from the field. Pierre Curie had found that in both diamagnetic and paramagnetic materials the strength of the induced magnetization rose with the external magnetic field, and in the case of paramagnetism also decreased with temperature. However, the strong remanent magnetization of a ferromagnet was very different. Curie envisaged there was some sort of interaction between the atoms that caused spontaneous magnetic alignment and resulted in an overall magnetization.

It had long been known that ferromagnets such as iron and lodestone lost their magnetization when heated in a fire. Curie showed the temperature at which this occurred—now known as the Curie temperature—was characteristic of the particular material, and that above this temperature ferromagnetic materials behaved much like paramagnetic ones.

Curie's work was published in 1895, just before the announcement by J.J. Thompson, Cavendish professor of physics at Cambridge, of the long anticipated discovery of the electron—the tiny, negatively charged particle present in electric currents and electric discharges. By 1905 atoms and electrons were all the rage in physics. Coulomb and Ampère had theorized that the basis of magnetism lay at the atomic level, but without a good picture of the atom they had been unable to develop the idea further. Now Langevin came up with an explanation of diamagnetism and paramagnetism by supposing that the electrons in an atom were themselves elementary magnetic dipoles. If the electronic dipoles in an atom cancelled one another out, a material made of such atoms was diamagnetic. If the elementary dipoles did not cancel one another out, the atom possessed a residual "dipole moment" and the material was paramagnetic. Using this model and the latest statistical theory, Langevin was able to account for Curie's experimental observations of paramagnetic behavior.

Weiss's contribution had been to explain the strong, stable magnetization of ferromagnetic materials by combining Curie's suggestion of interatomic interactions with Langevin's statistics. However, at the time (1907) there was still no good picture of the inner structure of the atom, and although it was suspected that the laws of physics might work a little differently within an atom, Weiss was in no position to even guess at the details. Instead, he assumed that the overall effect would be equivalent to an intense internal magnetic field, which he called the "molecular field," and which would act on each atom over and above any externally applied magnetic field. Using this idea, he predicted there should be a critical temperature above which a ferromagnetic material would become paramagnetic. In other words, his theory predicted the Curie temperature.

It was only after the studies of Curie, Langevin and Weiss that the New Zealand-born physicist Ernest Rutherford, experimenting at the University of Manchester's physics laboratory between 1909 and 1911, presented his famous picture of the atom as a tiny, central, positively charged nucleus surrounded by virtually empty space, occupied only by even tinier, negatively charged electrons. In 1920 Rutherford would name the nucleus's positively charged particles protons, and in 1932 James Chadwick identified neutrons, the uncharged particles that make up the remaining mass in the nucleus.

Hot on the heels of Rutherford's atom came an explosion of new ideas in physics. Scientists such as Niels Bohr, Erwin Schrödinger and Werner Heisenberg burst on to the scene, and quantum physics swept away many entrenched ideas of nineteenth-century classical physics. In particular, Heisenberg invoked the quantum mechanics of the electrons orbiting the nucleus of the iron atom to give a clear and lucid explanation of the interatomic interactions and molecular field that Curie and Weiss had deemed responsible for ferromagnetism.

French physicist Louis Néel, whose theoretical work showed that rocks can retain a stable record of the Earth's magnetic field, sometimes for hundreds of millions of years.

This, then, was the state of play in 1928 when Louis Néel joined Weiss's research group at Strasbourg. Immediately he noticed certain things that Weiss and Heisenberg had not, and which their theory could not explain. In particular some materials were not diamagnetic, paramagnetic or ferromagnetic: in the presence of a magnetic field they were as strongly magnetic as a ferromagnet, but removed from the field they showed no sign of a remanent magnetization. Néel hypothesized that interactions between atoms in these materials caused a spontaneous alignment in which the atomic dipoles perfectly cancelled; he coined the term "antiferromagnetic" to describe them.

In the years leading up to World War II, Néel began to study the magnetic properties of alloys. These are solid-state mixtures of two or more metals, in which the atoms are regularly arranged in a crystal structure. Magnetically, these alloys seemed to lie somewhere between ferromagnets and antiferromagnets, as if within them were two opposite but unequal magnetic alignments. For these Néel introduced a fifth class: ferrimagnetism, with the alloys becoming known as ferrites.

Néel had always had an underlying interest in geomagnetism. At one time in the mid 1930s he had considered taking up the old positions of Brunhes at Clermont-Ferrand but the opportunity had fallen through for lack of funding—alloys were a surer financial bet in hard economic times. In 1939 many of the Strasbourg group did relocate to Clermont-Ferrand, but Néel moved instead to Grenoble, where he would later establish a major international research institution, Centre d'Études Nucléaires de Grenoble (Grenoble Center for Nuclear Studies), and spend the remainder of his long and illustrious career. Early in the war he devised a method to demagnetize the hulls of ships and so protect them from the enemy's magnetic mines, and personally supervised its application to all major vessels in the French navy. As the war progressed and many other French cities were suffering miserably under German occupation, Néel took several Jewish scientists into his Grenoble laboratory. They would later prove to be invaluable collaborators.

The war over, Néel's interests returned to ferrimagnetism, particularly the "ferrites," which by then were being widely investigated with a view to producing stronger and more stable permanent magnets. This provided the basis for his work on naturally occurring magnetic minerals. The puzzle was how the magnetization gained by lava as it cooled in Earth's magnetic field came to be so strong and stable compared with the magnetization taken on by a rock that was simply left in a magnetic field of the same strength at room temperature.

Néel's first discovery was that he could explain the properties of many natural magnetic minerals—magnetite, for example—in the same way as he had his two-metal alloys. The crystal structure of magnetite meant that the iron ions were arranged in two networks of opposite but unequal magnetization. Like the ferrite alloys, magnetite was ferrimagnetic: it carried a strong, stable natural magnetization, and had a characteristic Curie temperature of 585°C.

Néel thought of a rock as containing a small percentage of ferrimagnetic grains, each uniformly magnetized. These grains were randomly distributed, and far enough apart that they did not interact magnetically. Each grain became magnetized when it first cooled through the Curie temperature of its mineral—about 500° to 600°C—but, as Néel showed, permanent, absolutely stable magnetization did not occur.

The stability of a grain's magnetization depended on the grain's size and shape, chemical composition, and the temperatures to which it was subsequently subjected. Néel's calculations showed, for example, that while iron grains up to twelve nanometers in diameter (a nanometer being a millionth of a millimeter) were unstable at room temperature, grains of 16 nanometers or more were, on the average, extremely stable. Collections of such grains might retain their magnetization for hundreds of years: even if the local magnetic field direction changed, the original magnetization would remain.

For natural minerals such as magnetite the figures were even more impressive. They clearly showed that fine-grained volcanic rocks and sediments might easily retain a stable magnetization for hundreds of millions of years—provided they remained at a sufficiently low temperature. If a rock was reheated close to its Curie temperature, the magnetization would become less stable, and likely to be affected by changes in the local magnetic field. The challenge for paleomagnetists would be to gain a deeper understanding of this trade-off between temperature and magnetic stability and to exploit this in unraveling the complexities of Earth's ancient magnetic field, hidden in its rocks.

Néel's discoveries laid the cornerstone of rock magnetism and paleomagnetism. Although only eight of his more than 200 publications, almost all in French, were directly concerned with the magnetization of rocks, the most important ones, which appeared between 1949 and 1952, were reviewed in English in 1955 and have since been cited in almost every work of paleomagnetism.

Rock magnetism was just one facet of a long career devoted to understanding magnetism and magnetic materials. When Néel received the Nobel Prize for Physics in 1970, it was noted that his pioneering work on the materials he called "ferrites," and his discovery of the properties of antiferromagnetism and ferrimagnetism, had been central to the fortunes of all the world's major electronic companies.

Poles Flipped,
Continents Adrift

*Measurements of the magnetism of English rocks indi-
cate that in the last one hundred and fifty million years
the country as a whole has turned through some forty
degrees.*

—*THE MANCHESTER GUARDIAN*, SEPTEMBER 9, 1954

By the late 1940s, research into the history of Earth's magnetism was
regaining momentum in Europe, and active research groups soon
sprang up in London, Manchester and Cambridge. Britain's first
contribution to the polarity reversal debate came when a New
Zealander, Edwin (Eddy) Robertson, arrived at London's Imperial
College wanting to test a new magnetometer he had designed for
field surveys. Robertson headed to the Isle of Mull in western Scotland, from where a swarm of thirty-million-year-old dykes radiate
south and east, rather like the Pilanesberg dykes in South Africa.

The British dykes are up to 180 miles in length, but do not outcrop continuously: long sections lie hidden beneath surface rocks

and soil. A magnetic survey is often useful in mapping such covered intrusions because they are usually strongly magnetized. To Robertson's great surprise, although his magnetometer recorded a strong signal it was exactly the opposite of what he had expected: the dykes were reversely magnetized. To be absolutely certain, he and John Bruckshaw, a colleague at Imperial College, collected samples to study in the laboratory. Their experiments confirmed that the dykes were indeed magnetized in the opposite direction to the magnetic field in the north of Britain.

In 1949 Horace Manley, a third member of the Imperial College team who had also become interested in the magnetization of natural materials, and of dykes in particular, wrote an amazingly perceptive review of the first half-century of paleomagnetism. Curiously, although he discussed early work on secular variation at length, he did not mention Brunhes' discovery of reversely magnetized rocks, nor the later studies of Mercanton or Matyama. However, he was in no doubt as to the cause of the "inversely" magnetized dykes:

> Sufficient experiments have now been made to allow only one plausible explanation of this "inverted" magnetization—that the Earth's magnetic field was itself reversed at the period when the rocks were formed.

There would, he forecast, soon be enough data to document the "history of the terrestrial magnetic field since rocks were solid at the Earth's surface."

Manley was well ahead of his time, but his article fired the imagination of a young Dutch scientist who was about to begin research at Cambridge University. Jan Hospers had been assigned the task of correlating Icelandic lava flows according to the strength of their magnetization. This would prove near impossible for all sorts of reasons, but the directions of the magnetizations caught

Hospers' attention. In two separate sequences of lava flows, he found that the younger flows, which he called the "gray phase," were magnetized in a normal direction, while the underlying older flows were reversed. In a third sequence there were even older lava flows that were again normally magnetized. Although each group of directions was quite scattered, it was obvious that the average directions were opposite one another: roughly parallel and anti-parallel to the present-day magnetic field.

Back at Cambridge, Hospers met up with two people who were to influence not just his studies but his future research in the entire fields of paleo- and geomagnetism. Keith Runcorn had recently arrived from Manchester, where he had just completed a PhD with the physicist Patrick Blackett. Blackett had won the Nobel Prize for Physics in 1948 for his development of the "cloud chamber" and his study of cosmic rays and so was best known for his work in this area of physics, but he also had an interest in Earth's magnetic field. He firmly believed that planetary magnetism was intrinsically related to rotation.

This had been the subject of Runcorn's PhD research, but he had failed to prove a link and was looking for fresh avenues of research. He immediately grasped the significance of Hospers' results and began to provide him with much-needed support, even though he was not Hospers' supervisor. In particular, he encouraged him to resample the Icelandic lava flows in order to replicate his results, and so validate the sequence of normal and reversed polarity intervals. In no time Runcorn became the nucleus of a young and energetic geomagnetism research group at Cambridge.

The second influence on Hospers was Ronald Aylmer Fisher. Fisher was a renowned statistician, a professor of genetics, and a Fellow of Gonville and Caius College, where he often dined with Runcorn. Several years earlier he had devised a statistical method of analyzing directional datasets, but his work still lay in a drawer, almost forgotten for want of a meaningful application. Now, during discussions with Runcorn over the dining table, Fisher recognized that Hospers' scattered paleomagnetic data offered a real-life use for his method. He got hold of the data and personally calculated the average directions and the angular equivalents of standard deviation and confidence limits, conferring on them the seal of scientific authenticity.

Hospers gave serious consideration to various possible causes of his reversely magnetized lava flows, including the idea that by some quirk of nature certain rocks might become magnetized opposite to the local magnetic field. At the end of the day he rejected all explanations but one: that the polarity of Earth's magnetic field had flipped, not once but at least twice—from normal to reversed and back to its current normal state.

From the rough estimates of age he had at his disposal, Hospers calculated that about half a million years had passed between polarity changes. However, he could find no intermediate directions. He therefore concluded that the reversal process itself must happen quite rapidly, perhaps taking less than ten thousand years.

Armed with Fisher's statistics, he went on to argue that the average magnetization directions of his lavas were just what you would expect from Gilbert's uniformly magnetized Earth or, equivalently, from a dipole at the center of the Earth aligned with the rotation axis—that is, a geocentric axial dipole. Statistically, the average direction of the normally magnetized lavas was indistinguishable from that of a geocentric dipole with its south pole "upwards," as is the case for the present-day field, while the average direction of the reversely magnetized lavas was indistinguishable from that of a geocentric dipole with its north pole "upwards."

This "geocentric axial dipole hypothesis"—the proposition that, over a long enough period of time, the average positions of the geomagnetic poles coincide with the geographic poles—was to become a central tenet of geomagnetism. However, its importance and implications were not fully appreciated immediately, any more than was the evidence for polarity reversals.

Manley had undoubtedly believed in geomagnetic polarity reversals, Hospers and Runcorn were equally convinced, but on the other side of the Atlantic, where the Carnegie Institution's Department of Terrestrial Magnetism had become something of a stronghold for secular variation studies, objections were mounting and a rival theory was brewing.

Before the war, scientists at the department had started a program to investigate whether soft, unconsolidated seabed sediments might carry continuous records of changes in Earth's magnetic field. After the war, a keen new research student called John Graham took up this work and extended it to older, consolidated sedimentary rocks. Graham kitted out a truck as a mobile field station and laboratory, and with a group of coworkers mounted several military-style expeditions across the United States to sample sedimentary rocks, some flat-lying and some deformed by folding.

Graham's primary goal was to assess the stability of the magnetization retained in these ancient sediments. Decades on, his legacy to paleomagnetism would be the field tests he devised to ascertain whether magnetization carried by a rock dated back to the time of its original formation, or had been reset during subsequent phases of deformation.

Graham's "fold test" was conceptually very simple. He imagined a flat-lying sedimentary rock that had been stably and uniformly magnetized at the time it was deposited. When such sediments were folded, he figured, the rocks on one side of the fold would be tilted one way, and those on the other would be tilted the other way. If the magnetization were stably locked into the rocks, it too would tilt so that it ended up in different orientations on different sides of the fold. Working backwards, Graham decided that if he could measure the angles through which the rocks had been tilted, he should be able to figure out by geometrical calculations the original direction of magnetization. If the clusters of magnetization directions from the two sides converged to one, he could conclude that the magnetization pre-dated the folding process, and could take this single direction as the original paleomagnetic field direction.

By the end of 1949 Graham had collected a set of data that, in other hands, could have led to all the groundbreaking discoveries later made by Hospers and his successors at Cambridge. But perhaps because he was conditioned by the conservative adherence to secular variation at the Carnegie Institution, he missed the clues and failed to follow the right leads. He came very close to the geocentric axial dipole hypothesis when he commented that most of his results from rocks spanning the past sixty million years clustered around geographic north, rather than the present field direction, but without the rigor of Fisher's statistics the claim lacked force. In any case, he used it more to argue that the field had been stable and had retained a normal polarity over that period of time than to question the difference.

In his fold-test studies, however, Graham had come up with one disconcerting piece of data. This involved the magnetization of the Silurian Rosehill Formation in Virginia, which dated back about 430 million years. After he had applied tilt corrections to samples from different parts of this fold, he found that the magnetization directions separated into two distinct clusters—one in a northwesterly and upward direction, and the other southeasterly and downward.

This was probably Graham's best example of a successful fold test, but how could he explain the resulting directions? There was not one cluster but two in roughly opposite directions, and neither was remotely close to the present magnetic field direction in Virginia, which was northerly and downward. Finally, after consulting his Carnegie Institute colleagues, Graham concluded that his rocks must have been magnetized during a period of anomalously high-amplitude secular variation. He wrote:

> My physicist colleagues . . . expect to find evidences . . . for stronger secular variation foci in remote epochs when the interior of the Earth may have been hotter and the disturbing current systems nearer to the surface . . . The corresponding current systems now appear to be roughly at 1200-kilometer depth.

However, Graham next considered another explanation—whether there was some fundamental way rocks could become magnetized in a direction opposite to the prevailing magnetic field. He consulted Louis Néel on the matter. Ernest Rutherford is reputed to have commented that if you ask a geologist to describe a stone he will conjure up the history of the entire Earth, but if you give the same stone to a physicist he will describe the minutest details of its atomic structure. Néel was the archetypal theoretical physicist: rather than credit the polarity-reversal hypothesis, he came up with several sub-microscopic processes by which such "self-reversal" might occur and described them in meticulous detail. Wholesale disruption of the Earth's inner constitution was apparently no longer required to explain reversely magnetized rocks.

All of Néel's self-reversal mechanisms required the combination of rather special physical conditions and unusual chemical compositions. How common were such conditions and compositions in nature? A logical test would be to take a sample of rock that has been found to carry a reversed natural magnetization, heat it above the Curie temperatures of its minerals, cool it in a known laboratory field and see which way it ends up magnetized. One of the first such experiments was carried out by a Japanese paleomagnetist, Takesi Nagata, in Tokyo and reported in *Nature* magazine in 1952. Nagata had found that samples of lava ejected from Mount Haruna, an active volcano in eastern Honshū, became magnetized in the opposite direction to the local field when cooled in temperatures between 440° and 250°C.

This was enough to reinforce the idea of self-reversal in many minds, including Graham's. In 1953, while Hospers was arguing that at least two polarity reversals had occurred in the past few million years, Graham wrote:

> . . . there is reason to doubt that inverse magnetization of rocks prove a reversal of the Earth's magnetic field; indeed, fuller understanding of the mechanism of their magnetization may be crucial in showing the constancy of the Earth's field.

And in 1954 he argued that:

> . . . during Paleozoic time, the Earth's magnetic field retained approximately its present orientation and, except for possible brief excursions, its present sense.

Graham had thrown the proverbial cat among the pigeons.

Back in Cambridge a frenzy of activity was underway. Edward (Ted) Irving, who was in the thick of it, would later describe these years as "fluid, even chaotic . . . No one was really in charge."

Inspired by Hospers' results, and imagining new and novel possibilities for the paleomagnetic method, Runcorn had hired Irving, a research student, and in 1951 the pair set out to sample the fine-grained, red Torridonian Sandstone of Scotland in search of records of ancient secular variation. On their way back to Cambridge they stopped off in Manchester to measure their samples using Blackett's now famous magnetometer. Through many unsuccessful attempts to verify Blackett's theory of the origin of the geomagnetic field, it had evolved into by far the most sensitive instrument available.

Like Graham's results, Irving and Runcorn's contained two opposite clusters of directions: northwest and upwards and southeast and downwards. Viewed either way, they were, according to Irving, "miles away from that expected if Scotland had not moved."

Irving and Runcorn were clearly in no doubt that the axis of the geomagnetic dipole had been always, on the average, aligned with the rotation axis. In other words, the average positions of the geomagnetic poles and geographic poles had always coincided.

Implicitly, they were assuming that Hospers' geocentric axial dipole hypothesis was good not just for the past few million years, but right back to the Cambrian—the past 500 million years. The fact that neither set of directions pointed in the geocentric axial dipole direction suggested that over the past 500 million years Scotland had rotated with respect to the poles, and carried the magnetization of the Torridonian Sandstone with it. This much had to be true, whether the opposite nature of the two clusters of directions was due to field-reversal or self-reversal.

All ideas of looking for secular variation went out of the window. Graham may have thought such results disconcerting, but to Irving and Runcorn they were the next exciting clue in the interlocked mysteries of geomagnetic reversals and continental drift. Was the long-awaited breakthrough just around the corner?

Before long the Cambridge group was augmented with the arrival of two new postgraduate students, Kenneth Creer and David Collinson. A Blackett-type magnetometer was built and, as

Edward (Ted) Irving, whose studies of the Scottish Torridonian
Sandstone led to confirmation that the polarity of the Earth's magnetic
field had "flipped" many times, and that the continents were drifting.
The Cambridge group of which he was a member also included Keith
Runcorn and Kenneth Creer.

Irving reported, "Ken raced through the Paleozoic"—sampling
and studying the Devonian Old Red Sandstone of South Wales,
the Permian Exeter volcanic rocks, and many other rock formations, all of which fell chronologically between Irving's Torridonian sandstone and Hospers' Tertiary lavas.

It came as no surprise that not one of the results pointed towards the geographic north pole. Creer later recalled that when their results were presented at a meeting of the Royal Society in May 1954:

> . . . one visitor asked for the latitude and longitude of the geomagnetic poles of Irving's Torridonian, which was our most thoroughly measured rock formation. In order to provide a quick answer to this question we had the Royal Society premises searched to find a world globe. Then to locate the ancient pole position quickly we improvised by using a piece of string . . . We found one of these paleopoles to be in the central Pacific and the opposite one in Ethiopia. But we did not have logarithmic tables nor a slide rule to be able to carry out numerical calculations there and then.

Later that summer, with Runcorn overseas, Creer was asked to take his place at the annual meeting of the British Association for the Advancement of Science in Oxford and speak about the work of the Cambridge group. By now Creer and Irving had calculated paleomagnetic poles ("paleopoles") for every one of their rock formations.

What did these paleopoles represent? Two hundred years earlier Gauss's mathematics had shown that a geocentric dipole accounted for by far the biggest part of Earth's magnetic field, and this had been amply confirmed by later calculations. Hospers had suggested that, averaged over hundreds of thousands of years, the net effect of the secular variation, particularly of the remaining non-dipole part of the field, came to nothing; this had led to his hypothesis that the time-averaged magnetic field since the Tertiary was equivalent to a geocentric axial dipole. Irving and Creer had calculated average magnetic field directions from each of their more ancient geological formations. These also represented long time-averages, but unlike Hospers' average directions they did not point towards the geographic pole.

Hence the question, where did the magnetic pole corresponding to the 500-million-year-old Torridonian paleomagnetic direction lie? Creer and Irving knew it would lie along the direction indicated by the declination, at an angular distance determined by the inclination: this was the basis of their globe and string construction. Only afterwards did they calculate the positions using geometry and spherical trigonometry.

When ordered chronologically, the paleopoles from the various geological formations showed an unmistakable trend: the youngest pole, calculated from Hospers' directions, coincided with the present geographic pole, but the older poles lay further and further away. Inspired by a diagram entitled "Paths of the north pole (relative to the continents)" that he found in Beno Gutenberg's 1951 book *The Internal Constitution of the Earth*, Creer plotted the path of the paleomagnetic poles from the pre-Cambrian to the present.

Creer had his first "polar wander path" ready in time for the meeting of the British Association for the Advancement of Science in September 1954. He would later recall that those present were "receptive although not at all enthusiastic about continental drift or polar wander." However, media coverage of the meeting told a different story: *The Times*, *The Manchester Guardian* and even *Time* magazine all ran excited articles.

Despite adopting the name "polar wander path," Creer and his coworkers did not really believe that either the paleomagnetic or geographic poles had wandered. The geocentric axial dipole hypothesis implied that they coincided, and physical realities made wholesale movement of the rotation axis all but impossible.

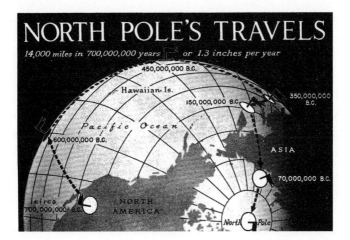

NORTH POLE'S TRAVELS

14,000 miles in 700,000,000 years ⬛ *or 1.3 inches per year*

450,000,000 B.C.

Hawaiian· Is.

150,000,000 B.C.

350,000,000 B.C.

Pacific Ocean

600,000,000 B.C.

ASIA

70,000,000 B.C.

circa 700,000,000 B.C.

NORTH AMERICA

North Pole

Time magazine's 1954 depiction of the apparent movement of the north pole, as seen from Britain, from 700 million years BC to the present—as deduced by the Cambridge paleomagnetism group of Creer, Irving and Runcorn. This work led to the confirmation of continental drift and the theory of plate tectonics.

So what exactly was wandering? The only other option was that the rocks, and hence the land mass to which they were fixed, were moving with respect to the poles. Here at last, then, was evidence that the British Isles, and likely as not the whole continent of Europe, were drifting: at the time the Torridonian Sandstone was laid down it was Britain, not the pole, which had sat close to the equator and sweltered. The polar wander path was really a record of continental drift.

Creer's polar wander path confirmed that the British Isles had moved significantly over the globe in the 600 million years since the Precambrian period. However, it did not answer the question of whether America and Europe had ever been closer together, as the fit of the coastlines suggested. Had the whole crust of the Earth rotated over the inner parts of the globe? Or had different parts of the crust moved independently?

This could be tested. If the whole crust had moved intact, the polar wander paths of different continents would overlie one another. If parts of the crust had moved independently, the paths would have evolved differently as the continents drifted with respect to each other. To solve the mystery, what were needed were polar wander paths for all the continents. Leaving Creer and Irving to tidy up at home, Runcorn set off to the United States to fire up paleomagnetists there and forge collaborations with them.

By 1957, the Cambridge trio had amassed enough data to construct a polar wander path for North America, as well as to improve the British path. The comparison was breathtaking: for the past few tens of millions of years the two polar wander paths had coincided, but before this recent and relatively short period of Earth's history the two had been offset by an almost constant longitude difference of about twenty-five degrees. To see this effect, have a look at a world map. If you move South America east by about twenty-five degrees, it slips right into the big bend in the West African coastline, give or take a margin of continental shelf.

The coincidence was too good not to be true. In fact, it was difficult to explain a systematic offset like this in any other way. Creer and company certainly tried: they checked for problems with the magnetic recording process, for the presence of secondary components of magnetization, and for errors in the assignation of ages. However, in each case it was hard to see why all the rocks on one continent should be affected, while none on the other were.

As they would state in *Philosophical Transactions of the Royal Society of London,* they kept coming back to the conclusion that:

> . . . in pre-Jurassic times, Europe and North America were several
> thousand miles closer than they are today. Such a displacement

. . . is described in the literature as continental displacement or drift. In view of the unsatisfactory nature of the other explanations, it is postulated that in Paleozoic and early Mesozoic times Europe and North America were very much closer together, and at some time prior to the mid-Tertiary they moved apart to their present positions.

Here was evidence that for most of Earth's geological history Europe and North America had been joined. They had moved apart only "at some time prior to the mid-Tertiary." And since the polar wander paths were recorded in the magnetization of the rocks, the paths too had separated by an amount equal to the eventual separation of the continents.

By 1957 when this was published several members of the Cambridge group had moved on and taken paleomagnetic research to new shores, but one thing was clear: the group was convinced that polarity reversals were real, as was the geocentric dipole nature of Earth's magnetic field and the overall correspondence of the geomagnetic and geographical axes—the so-called geocentric axial dipole hypothesis. Their statement was unequivocal:

> Because the magnetic poles reverse there are two possible positions for the north geographical pole for any geological period.

The question now, of course, was what caused such reversals. Runcorn, who, according to his students, "knew no geology" and "was best kept out of the laboratory," had emerged as the group's theoretician. Since his early experiences with Blackett, he had become a dedicated convert to the theory that Earth's magnetic field was generated by dynamo action. Unlike Graham, who seemed happy to hypothesize that electric currents were flowing in the mantle, he firmly believed that the field-generating currents would prove to be in Earth's liquid outer core.

Accordingly, he reasoned that two things were the case. First, the coincidence of the geomagnetic and geographical axes was a natural consequence of Earth's rotation acting on the fluid core. Second, polarity reversals could be brought about by "minor changes in the pattern of convection in the core." And so:

> The case for special causes of reversals of magnetization is not strong when reversals of the main field give a simple and general explanation for the varied phenomena observed.

The Cambridge group had also noted that polarity reversals and polar wander operated on different timescales. Reversals appeared to be relatively frequent—Hospers had suggested they occurred at intervals of about half a million years—while polar wander was very slow and gradual. On the basis of this, the group argued that the two phenomena must come about through different processes. But if reversals originated in fluid motions in Earth's core, what caused polar wander?

Keith Runcorn continued to be bugged by such questions, and began to consider mechanisms for the generation of Earth's magnetic field and for continental drift. Meanwhile, Ted Irving—who had now moved to Canberra and set up his own laboratory at the Australian National University—and Ken Creer addressed the obvious need to construct polar wander paths (or apparent polar wander paths, as they soon became known) for the other continents.

While British scientists were now convinced about polarity reversals, there were still doubters. As late as 1963, when the question was put to the vote at a meeting of twenty-eight eminent paleomagnetists in Munich, four reckoned that all instances of reversed magnetization were caused by self-reversal, fourteen thought that although self-reversal did occur there was sufficient evidence to believe that some records represented field reversal, and ten could not make up their minds. Not one voted unerringly for polarity reversals.

The problem was that while there was evidence for self-reversal—the Haruna lavas had a reversed natural magnetization and clearly showed self-reversing properties in the laboratory— there was no such direct proof of field reversal. Rather, the evidence was circumstantial. To unambiguously reject the field reversal theory, it would be necessary to demonstrate that all rocks carrying reversed magnetizations were capable of self-reversal, while none of the normally magnetized ones were. The accumulating evidence indicated that roughly half of all rocks studied were reversely magnetized, but examples of the peculiar minerals and crystal structures necessary for self-reversal remained extremely few and far between.

Meanwhile, indirect evidence for field reversals was gaining strength. First, although dating rocks was not easy—for sediments, it was traditionally based on identification of fossils, and for lavas on estimates of the frequency of eruptions—there was growing evidence that all across the world rocks of the same age recorded the same polarity. A timescale of geomagnetic polarity reversals was beginning to emerge. It was hard to come up with a reason for self-reversing and non-self-reversing rocks to occur in chronologically controlled sequences.

Secondly, there were a growing number of cases in which the directions of magnetization of lavas and the baked clays underlying them were the same. A rather special example was reported in 1962 by Rod Wilson, a geologist at Liverpool University's Department of Earth Sciences. Wilson had found an example of sediment that had been baked not once, but twice. It had first been overlain by a lava flow and its upper margin remagnetized in a reversed direction, which was also recorded in the lava. Subsequently, a dyke had pushed up through both the sediment and the lava. The dyke was also reversely magnetized, but in a direction that, because of secular variation, differed measurably from that of the lava. The parts of the sediment immediately adjacent to the dyke had been heated by it and remagnetized. Some parts that were close to both the lava and the dyke had been heated and remagnetized twice, but not necessarily to the same extent each time.

By carefully heating his samples to successively increasing temperatures, cooling them in the absence of a magnetic field, and then measuring their remaining magnetization, Wilson had been able to separate out the components of magnetization carried by grains of differing stability. This, in principle, would unravel the entire magnetic history of his rocks.

This study was too much for almost all the remaining adherents of self-reversal. Even if one in a thousand rocks was capable of self-reversal, the chance of finding one self-reversing sediment, one self-reversing lava and one self-reversing dyke in such close proximity was one in a thousand million—not quite impossible, but very nearly.

The case for polarity reversals was now almost proven. In just a few years, even more compelling evidence was to appear for both polarity reversals and continental drift—and from another surprising source.

The Story on the Seafloor

From this moment I never had any doubts about the concept. It locked three theories together in a mutually supporting way: continental drift, ocean floor spreading, and the periodic reversing of the Earth's field.

—Lawrence Morley, 1986

Attempts to survey the depths of the world's oceans go back to the 1850s, when the first transatlantic telegraph cable was laid between Ireland and Newfoundland. Contrary to previous assumptions that the seafloor must be flat and featureless, it was found the cable had to traverse a range of underwater mountains that rose 2000 meters above the seabed. These original bathymetric surveys were conducted by dropping a lead sinker down to the seabed on the end of a sounding line. Such work was laborious, tedious and prone to errors. To chart the entire ocean floor in this way would have been totally impractical. Clearly a more efficient method was needed.

During World Wars I and II the Allied nations were terrorized by German submarines. Winston Churchill is said to have admitted that the only thing that really frightened him was the U-boat threat. Not surprisingly, then, considerable British and American efforts went into devising methods of detecting the steel submarine hulls—efforts that resulted first in ultrasonic echo-sounding techniques, and later in airborne and rugged seagoing magnetometers.

Sonar detection technology evolved during World War I from the pioneering work of, among others, a French physicist, Paul Langevin. This technology works by continuously timing the reflection of high-frequency sound waves from the seafloor or other objects: since the speed at which the waves travel in water is known, the distance to the seabed or object can be calculated. By the 1950s a detailed picture of the ocean floors was emerging from sonar surveys. It was discovered that the mid-Atlantic Ridge, traversed by the first telegraph cable, was just part of a 40,000-kilometer-long chain of underwater mountains running around the globe, rather like the seam around a tennis ball. The chain snakes down the Atlantic, through the southern oceans, and up the eastern part of the Pacific. The average height of the mountains above the surrounding sea floor is about 4500 meters, and their flanks spread out to a width of 1000 kilometers or more.

The ridges—or rises as they are sometimes called—had several notable features, clues to the mechanism of continental drift. First, most ridges had a distinct trough or valley along the top. Second, close to these ridges scientists found an unusually high rate of heat flowing up from the Earth's interior. Third, when drill cores were retrieved from the ocean floor the covering layer of sediment was found to be much thinner than expected, particularly near the ridges.

Fourth, and most surprisingly, the whole of the seafloor was found to be geologically very young. Radiometric methods of dating rocks were in their infancy, but early results showed that whereas the oldest continental rocks were several billion years old, nothing on the seafloor was older than about 150 million years: the oldest seafloor had existed for only a tiny fraction of the lifetime of the Earth.

And lastly, the mid-ocean ridges frequently jinked to the right or left along well-defined lines, or fracture zones, also delineated by submarine mountains.

Furthermore, in other places far removed from the ridges, most notably around the margins of the Pacific Ocean, there were deep trenches that also extended many thousands of kilometers. Seismologists were beginning to notice that a high proportion of the earthquakes they recorded, as well as much of the world's volcanic activity, was concentrated along these ocean trenches and ridges. Clearly both played an important role in the process of continental drift—but how?

By 1962 two men had come up with the seed of an explanation. Harold Hess was a Princeton University geology professor who had served in the United States navy during World War II and Robert Dietz was a scientist in the navy's electronics laboratory. The anomalously high heat flow at the mid-ocean ridges had set both of them wondering about the source of that heat deep within the Earth, and they had recalled the idea of convection—heat carried by the movement of fluid material.

The notion of convection in the Earth went back to the 1830s, when Cambridge mathematician William Hopkins had considered the possibility of fluid layers within the planet. Hopkins was thinking of the sort of rapid convective circulation commonly seen in a pot of soup on a stove. Hot material at the bottom of the pot expands, becomes buoyant with respect to the cooler, denser fluid above, and therefore wells up. When it reaches the surface it loses heat and flows sideways, until it becomes cool enough and dense enough to sink again, thus completing a cycle, or "convection cell."

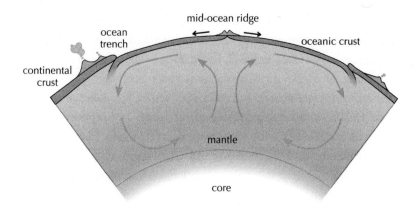

Convection in Earth's interior. As the mantle is not quite solid, hot material rises up under mid-ocean ridges and cooler material sinks at ocean trenches. Harold Hess and Robert Dietz suggested that, carried along by this convection, the oceanic crust spreads symmetrically away from the ocean ridges and is eventually drawn underneath the abutting continental crust.

In 1944 the idea of very slow convection in a much more viscous mantle had been revived by a Scottish geologist, Arthur Holmes, in an early attempt to explain continental drift. But at the time Holmes had found himself a lone "Drifter," a small voice in a crowd of "Fixists."

Hess and Dietz now suggested that the upwelling from this slow mantle convection reached the Earth's surface at the mid-ocean ridges. Where the measured heat flow was highest, they said, magma must be rising through the mantle, erupting and forming new crust. As new crust formed at the ridges, older crust moved away in both directions, riding on the thermally driven convective motion of the upper mantle. Furthermore, beyond the edges of the oceans, the continental crust was also riding high on the denser mantle.

Finally, the formation of new crust at the ocean ridges was being accompanied by the sucking down of old crust back into the mantle at the ocean trenches. Earthquake locations, Hess and Dietz suggested, delineated a downward-sloping surface at each ocean trench, on which the old seafloor slid beneath the abutting continent, creating the downgoing part of the mantle convection cell. Dietz introduced the term "seafloor spreading," which has been used ever since.

Although Hess and Dietz's ideas were, by the men's own admission, simply intuitive starting points, they fitted well with the geophysical observations and provided a credible mechanism for continental drift. Thirty years earlier Harold Jeffreys had complained that Alfred Wegener's theory of continental drift would have the continents "plough" through Earth's rigid lithosphere— the crust plus the uppermost mantle. He had finally been silenced.

The next breakthrough, which would tie geomagnetism into the seafloor spreading story, followed the development and deployment of the fluxgate magnetometer in World War II. The first fluxgate had been constructed in Germany in 1936, but it was its use by the Allies that would lead to its post-war geophysical applications.

The fluxgate magnetometer is an electromagnetic instrument that capitalizes on the strong, easily induced magnetization of ferrite materials, such as those Néel had been developing. Flown at low level over the sea, it had proved Churchill's best means of detecting the dreaded U-boats. After the war, fluxgate sensors began to be routinely deployed in watertight streamlined "fish," which were towed some 200 to 300 meters behind survey ships.

From the late 1950s, they were gradually superseded at sea by another new class of instrument, proton precession magnetometers, which use the principle of nuclear magnetic resonance to measure the intensity of the prevailing magnetic field. These were well suited for use on boats: their operation was independent of orientation, something always difficult to establish with precision at sea, and they did not require the regular calibration checks essential for fluxgates.

Because of these advances, high-resolution magnetic field intensity measurements began to be carried out around the oceans of the world alongside detailed bathymetric profiling. At first, measurements were made along lines or profiles, but with the beginning of systematic mapping—surveys consisting of series of parallel profiles—startling patterns began to appear. Charts of "magnetic anomalies" revealed long, alternating bands of high and low field-strength running parallel to the ridge systems, with steep gradients in between.

In the first detailed survey, which took place off the west coast of North America, this pattern was found to repeat again and again, over huge areas of seafloor. The regular alternations were completely different from the almost random pattern of magnetic anomalies that scientists were used to seeing in aeromagnetic surveys over land, in which a magnetometer was carried by an airplane. When positive anomalies were colored black and negative anomalies white, the marine magnetic charts resembled the stripes on a zebra's back.

The common opinion among geomagnetists was that the patterns were the result of magnetization induced in rocks on the seafloor by Earth's magnetic field. Black (positive) anomalies were supposed to indicate highly susceptible, easily magnetized rocks, and white more weakly magnetizable rocks. But why did strongly and weakly susceptible rocks alternate in this way? Why should the magnetic composition of seafloor rocks change so rapidly and regularly? And why should the strongest anomalies always occur right over the mid-ocean ridges, broken only by sharp offsets that mirrored the fracture zones?

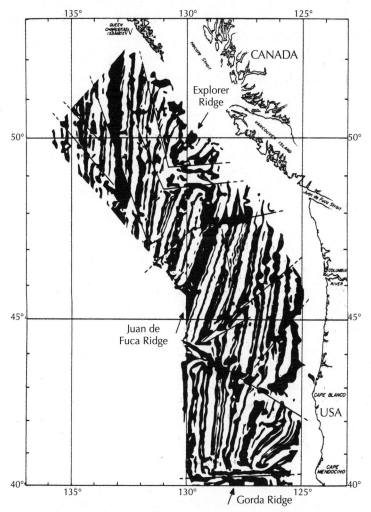

A marine magnetic survey off the west coast of the United States and Canada, originally published in 1961. Black shading shows areas where the magnetic field is stronger than expected, and white shading where it is weaker. The significance of the striped patterns was not recognized at the time. Only later were the segments of the East Pacific Rise—the Gorda, Juan de Fuca and Explorer Ridges—recognized, and the magnetic anomalies interpreted in terms of seafloor spreading and reversals of Earth's magnetic field.

In the summer of 1962, Drummond Matthews, a marine geo-physicist returned to Cambridge University from a scientific cruise in the Indian Ocean to find his new research student, Fred Vine, eagerly awaiting instructions. Matthews set Vine to work on the magnetic, bathymetric and gravity data he had just collected over the Carlsberg Ridge of the northwest Indian Ocean. At the same time, on the other side of the Atlantic, Lawrence Morley, chief of the Canadian Geological Survey's Geophysics Division, was becoming increasingly distracted by the extraordinary marine anomaly records coming in from the east Pacific Ocean.

Vine, Matthews and Morley were all familiar with the reversed remanent magnetization now found increasingly frequently in rocks. All three were adherents of the field-reversal theory, so not surprisingly they put two and two together at the same time. Seafloor spreading, combined with geomagnetic polarity reversals and remanent (rather than induced) magnetization could, they realized, explain the seafloor stripes.

Vine and Matthews wrote to *Nature* magazine putting forward this theory—and so did Morley. In a faux pas of the scientific refereeing system, Vine and Matthews' letter appeared in *Nature* on September 7, 1963 but Morley's was turned down. Morley's manuscript was then rejected again, this time by the *Journal of Geophysical Research* with the comment "interesting" but "more appropriately discussed at a cocktail party than published in a serious scientific journal." It was over a year before his paper was eventually published by the Royal Society of Canada, and later still that the theory became known as the Vine-Matthews-Morley hypothesis.

This hypothesis proposed that magma erupting at a mid-ocean ridge would, when it cooled below the Curie point of its magnetic minerals, become stably magnetized parallel to Earth's magnetic field. It would then begin to spread in both directions on the

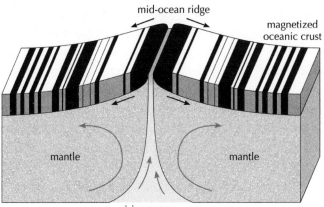

mid-ocean ridge

magnetized oceanic crust

mantle

mantle

rising magma

As the seafloor spreads away from mid-ocean ridges, new magma rises and solidifies. As this magma cools it becomes magnetized in the direction of the prevailing magnetic field: the seafloor therefore becomes a record of polarity flips. In this "barcode," black represents normal polarity, when compasses would have pointed north, and white reversed polarity, when they would have pointed south.

conveyor belt of the mantle convection cell. Closest to the ridge you could expect to find rock magnetized in the present "normal" field direction, but further away you could expect to find older rock, which had erupted and cooled before the last polarity reversal, magnetized in the opposite direction. The seafloor would, therefore, be acting like a double tape recorder, advancing in both directions from the ridge, producing strips of rock magnetized alternately in "normal" and "reversed" directions, symmetrically arranged about the ridge; the width of each strip would be determined by the length of the polarity interval and the rate of seafloor spreading.

Morley's reasoning was purely conceptual, but Vine and Matthews tested theirs by calculating the anomalous magnetic field that such alternately magnetized blocks would produce, and comparing the results with the data they had collected from the Carlsberg Ridge in 1962. Lucky enough to have access to a computer at the Cambridge University Mathematical Laboratory, they programed it to model magnetic anomaly patterns corresponding to the Pacific Rise at a magnetic latitude of about 40°, the mid-Atlantic Ridge at a latitude of about 47°, and the Carlsberg Ridge. Their models turned out to be a far better fit with the observations than models based on the earlier idea of normal induced magnetizations of varying strength.

In 1963, when Vine and Matthews' results were published, the field-reversal theory was still controversial—as seen in Munich, when half of the twenty-eight assembled "experts" had either rejected the idea or been unprepared to commit themselves to it. Equally, seafloor spreading was still an intuitive theory not fully proven. Vine and Matthews' original calculations had sidestepped details such as the rate of spreading and the dates of polarity reversals by assuming twenty-kilometer wide strips of seafloor that were alternately normally and reversely magnetized.

Morley had gone on to do some rough calculations based on early estimates of the frequency of reversals and the length of magnetic anomaly patterns, and had shown the concept was consistent with a rate of seafloor spreading, or mantle convection, of a few centimeters a year. This was a credible figure, but it still did not constitute hard proof.

Unraveling the Record

The entire history of the ocean basins in terms of ocean-floor spreading is contained frozen into the oceanic crust.

—Fred Vine, 1966

The Vine-Matthews-Morley concept was slow to catch on among marine geophysicists, and received little immediate attention from other sectors of the Earth Sciences community. In the next few years, the focus of United States marine geophysical work shifted from the west coast to Columbia University's Lamont Geological Observatory, now the Lamont-Doherty Earth Observatory, in New Jersey, where Jim Heirtzler headed up operations. Heirtzler was a traditionalist, and would prove hard to convert to the theories of seafloor-spreading and polarity reversals.

Meantime, quite independently, the race to validate the field-reversal theory by establishing a global timescale for polarity reversals was in full swing among paleomagnetists. Earlier attempts had been hampered for want of a precise method of dating rocks. Traditional fossil-based dating methods were adequate for defining geological periods, which typically lasted hundreds of millions of years, but the speed of evolution and migration of species was too slow to resolve polarity intervals that lasted only a few million years at most.

During the 1950s, new dating techniques had been developed. These involved measuring the radioactive decay of certain isotopes of naturally occurring elements. At the beginning, the methods focused on isotopes that decayed very slowly; they had been applied to some of the oldest rocks and meteorite fragments to investigate the age of the Earth and of the solar system itself.

One such method involved the decay of an isotope of potassium, ^{40}K. In ^{40}K the nucleus of the atom has twenty-one neutrons in addition to the standard nineteen protons. Most naturally occurring potassium is ^{39}K, in which atoms have twenty neutrons; only about one-hundredth of a percent of naturally occurring potassium is ^{40}K. This tiny but incredibly evenly distributed fraction is radioactive: in 1250 million years half of the ^{40}K nuclei in any sample will decay—89.5 percent to calcium (^{40}Ca) and 10.5 percent to argon (^{40}Ar). Potassium–argon dating relied on comparing the amount of ^{40}Ar that had accumulated in a rock since its formation with the amount of ^{40}K remaining. This ratio, together with an accurate knowledge of the half-life and details of the decay process, could be used to calculate the time since the minerals in the rock crystallized.

Although radiometric dating was a vast improvement on fossil-based methods, early potassium–argon age estimates still lacked enough precision. Uncertainties and errors centered on possible contamination by the small amount of argon—about one percent—present in Earth's atmosphere, and the need to eliminate this from the mass spectrometers used to measure isotope ratios.

In the early 1960s, the pioneering laboratory in potassium–argon dating was at the University of California, Berkeley, where geologists Brent Dalrymple and Sherman Grommé and petrologist Ian McDougall, on an overseas fellowship from Australia's Commonwealth Scientific and Industrial Research Organization, were engaged in graduate and postgraduate work. Meanwhile, at the United States Geological Survey in nearby Menlo Park, geophysicists Allan Cox and Richard Doell were also forging ahead in pursuit of a timescale for geomagnetic reversals.

In 1961 McDougall returned to the Australian National University in Canberra, stopping off in Hawaii en route to sample sequences of lava flows. In Canberra he set up his own potassium–argon dating facility, and teamed up with Ted Irving and Don Tarling, a new graduate student from Britain, to tackle the polarity timescale problem from Down Under.

The race was on. Over the next few years the lead bounced back and forth across the world as first the Californians, then the Australians, published new results, new age estimates and ever more evidence for sequences of field reversals. The story unfolded in a series of letters and papers, published mostly in *Nature* and *Science*, that described successive versions of what came to be called the Geomagnetic Polarity Time Scale, GPTS.

In 1963 Cox and Doell acquired samples from six Californian lava flows that Dalrymple had been dating using his high-precision, argon-free, mass spectrometer. They tested the suitability of the samples for paleomagnetic work and, when satisfied, measured the directions of their remanent magnetization. These six results, together with three from Italian lava flows published earlier by Martin Rutten, professor of geology at the University of Utrecht in The Netherlands, would form the basis of the first accurately dated polarity timescale.

Uncertainties in the estimates of age were about five percent, small enough that the samples could be unambiguously ordered. Three normally magnetized flows were dated at between 0 and 0.98 million years, three reversed flows at between 0.99 and 1.69 million years, and another three normal flows at between 2.4 and 3.2 million years. Although Cox and his colleagues were careful to point out that other scenarios were possible, their results supported the suggestion that first Matuyama and then Hospers had made— namely that the last reversed-to-normal polarity change had occurred sometime in the early Quaternary period, about a million years ago, and there had also been an earlier period of normal polarity in the late Tertiary period. They provisionally labeled the present period of normal polarity N1, the earlier one N2, and the intervening period of reversed polarity R1, and tentatively suggested that reversals might occur regularly, roughly every one million years.

Later in 1963, McDougall and Tarling announced that the dates and polarities of their Hawaiian samples were consistent with the results of Cox, Doell and Dalrymple, except for one reversely magnetized flow that they had dated at 2.95 million years, with an uncertainty of 0.06 million years, placing it in the middle of the late Tertiary normal interval N2.

The following year any semblance of tidy regularity disappeared for good when Cox and his colleagues published their next version of the timescale. Two short "events" had broken into the long polarity "epochs," as well as an earlier reversed interval, R2. At Mammoth Lakes in California, they had found a reversely magnetized lava flow, which they dated at about 3.1 million years. Uncertainties in their estimate of age meant they could not be sure whether their Mammoth Lakes result matched McDougall and Tarling's rogue one, or whether there had been two separate events: for some time a question mark would remain over the Mammoth event.

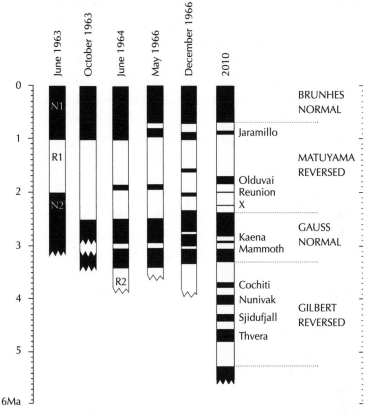

Early steps in unraveling the history of geomagnetic polarity reversals. In 1963 Allan Cox and his team suggested there had been alternating intervals of normal and reversed polarity, with each spanning about one million years. Later studies resolved shorter polarity "events" within Cox's original "epochs" and improved the ages of the polarity boundaries. By 1966 the timescale bore little semblance of regularity. On the right is a modern version of the geomagnetic polarity timescale for the past six million years, with the main epochs named after famous geomagnetists, and the shorter events after the locations where they were first identified.

The 1964 version of the timescale also included a famous "event" that had first been noticed in Olduvai Gorge in Tanganyika (today's Tanzania). The excavations at this East African location were well known for their early hominid fossils, for which accurate and precise dating was important. Between the fossil-bearing sediments, a volcanic "tuff" layer, originally dated at 1.9 million years, was found to be normally magnetized when "it should have been reversed," according to Sherman Grommé and Richard Hay who reported it.

The possibility of self-reversal was never far from the minds of the early paleomagnetists, and at first it seemed that the Olduvai Gorge tuff might be another rare example to rank alongside Nagata's Haruna lava. Cox and his coworkers now instigated a strict regime of tests to validate their results. To rule out self-reversal, they heated their samples in the absence of a magnetic field and cooled them again in the laboratory field. To identify and remove unstable secondary magnetizations, they progressively demagnetized the samples in alternating magnetic fields of increasing strength. To check for consistency, wherever possible they sampled different parts of a flow and baked contacts.

By the end they had found no evidence of self-reversing rocks and no flows with ambiguous results. Each appeared to indicate a field of either normal or reversed polarity: the samples from Olduvai Gorge were unequivocally normal. Cox and his colleagues were happier still when they found a second record from a volcanic island in the Bering Sea that corroborated the Olduvai event.

By now the polarity timescale was made up of results from North America, Hawaii, Europe and Africa, ordered chronologically and interleaved. The regular polarity epochs were punctuated with just two short events: the Olduvai, 1.9 million years ago, and the Mammoth, about three million years ago.

The consistency of the results from three continents, together with the failure to find any evidence of self-reversal, provided the proof needed to convince almost all remaining doubters that the polarity of Earth's main magnetic field had indeed reversed several times in the recent history of the planet. Cox and his team surreptitiously slipped in the suggestion that the epochs be named after pioneers in the field of geomagnetism, and so N1, R1, N2 and R2 became known respectively as the Brunhes, Matuyama, Gauss and Gilbert, names that have since survived all subsequent attempts to regularize them.

Cambridge University has always been something of an intellectual crossroads, and it was a fortuitous meeting of minds there in early 1965 that would give Fred Vine the incentive to develop the Vine-Matthews-Morley hypothesis to fruition. For eighteen months the idea had languished. Vine had been at a loss as to how to continue. He had been unable to find good magnetic anomaly records over ocean ridges that were believed to be spreading centers, and no reliable geomagnetic polarity timescale had been available for comparison. The patterns of magnetic anomalies recorded in the northeast Pacific were tantalizing, but it was not obvious where, or even whether, an active spreading ridge lay at their heart.

When Tuzo Wilson, a geophysicist from the University of Toronto, and Princeton's Harold Hess arrived in Cambridge, both on sabbatical leave, almost everyone was on vacation. Fred Vine was holding the fort at the Department of Geodesy and Geophysics, and coffee-time conversation between the three naturally turned to seafloor spreading and associated questions.

Vine was to recall later that Harold Hess gave him a much-needed boost of confidence. "One of the first things he said was that he thought the Vine-Matthews hypothesis was a fantastic idea. No one had ever said that to me before," Vine would tell William Glen, a geologist and science historian who researched the story for his 1982 book *The Road to Jaramillo*.

Tuzo Wilson would provide the last crucial piece of the puzzle. It was now known that new crust formed along ocean ridges, and that old crust returned to the mantle at trenches. However, something more was needed to make the whole process work on a spherical globe. Wilson postulated that Earth's surface was a series of mobile plates, bounded by the ridges and trenches, and that between them were what he called "transform faults" along which the plates slid sideways, parallel to the faults.

These were a new-found type of fault, and the direction of motion along them would prove whether or not Wilson's concept of "plate tectonics" was correct. But what interested Vine was Wilson's suggestion that there was a whole series of these transform faults in the eastern Pacific Ocean, offsetting segments of the East Pacific Rise. In particular, the Rise underwent a major offset along the San Andreas transform fault, and in the northeast Pacific it consisted of three relatively short ridge segments, each offset from the other by more transform faults: the Juan de Fuca Ridge, named by Vine and Wilson, and the later-named Gorda and Explorer Ridges.

Wilson's breakthrough sent Vine straight back to his computer. He now knew exactly where to place the spreading centers in his seafloor models. To calibrate for age he used the latest geomagnetic polarity timescale of Cox and his colleagues—major reversals at 1.0, 2.5 and 3.4 million years ago and short events at 1.9 (the Olduvai) and possibly 3.0 (the Mammoth) million years ago.

To start with, Vine computed the magnetic anomaly sequences you would expect if the seafloor were spreading at the rates of one centimeter a year and two centimeters a year. The first was too slow. The second, although too fast, did reproduce many of the details that were observed in the actual profiles. By judiciously adjusting the boundaries of the strips of normally and reversely magnetized seafloor in his model, he came up with a "best fit." However, to reconcile his model with Cox's timescale he had to accept significant variations in the rate of spreading, particularly around one million years ago. Although he tried to justify these variations, he was clearly not happy with them. However, he was not confident enough to suggest there might be a problem with the timescale.

Reading Vine's paper "Spreading of the Ocean Floor: New Evidence," published in *Science* in December 1966, you can imagine him cursing his prior lack of confidence. In the new version of the polarity timescale they released that year, Doell and Dalrymple had revised the age of the Matuyama–Brunhes boundary from 1.0 million to 0.7 million years, and added a new normal polarity event, the Jaramillo, near the top of the now expanded Matuyama Epoch, at 0.9 million years.

In effect, an interval of reversed polarity had been inserted into the old scheme between 0.9 and 0.7 million years ago. With this addition, Vine's model for the Juan de Fuca Ridge system matched the geomagnetic polarity timescale with an almost constant spreading rate of 2.9 centimeters per year.

By now, Vine had left Cambridge and was a member of the teaching staff at Princeton University. His paper was a summary of his landmark PhD thesis—but it also drew on a remarkable new marine magnetic record from the South Pacific Ocean. From September to November 1965 the USS *Eltanin*, a United States naval ship operating under the auspices of the National Science Foundation, had collected magnetic and seismic data over the Pacific–Antarctic Ridge system. In particular, it had made measurements along two lines, *Eltanin*-20 and -21, to complement a previously measured line, *Eltanin*-19. In *The Road to Jaramillo*, William Glen writes of Vine's excitement at spotting these *Eltanin* records when he visited Jim Heirtzler at the Lamont laboratory early in 1966. He instantly recognized the pattern of wiggles: it had been imprinted on his memory during long hours of work on the Juan de Fuca records.

Here was inescapable, independent confirmation that the seafloor carried a record of the sequence of geomagnetic polarity reversals.

Vine requested and was given a copy of the *Eltanin*-19 profile by Heirtzler—unbeknown to the two young research students, Walter Pitman III and Ellen Herron, who had taken part in the *Eltanin* cruise. The students had just finished laboriously processing the *Eltanin* magnetic data, and had immediately realized its importance. Naturally, they were furious that at that very moment their supervisor had handed over their treasured discovery to their rival. Only after much heated correspondence was an agreement eventually hammered out with the publishers of *Science*: Pitman and Heirtzler's presentation of the *Eltanin* data and their interpretation would appear two weeks before Vine's "Spreading of the Ocean Floor: New Evidence."

Vine's paper included seafloor data from around the world. It all told the same story. At last scientists could unravel Earth's magnetic history millions of years back into geological time, and move the debate on the source of this magnetism to an exciting new level.

Vine began by looking at a detailed low-altitude aeromagnetic survey of the Reykjanes Ridge, the part of the mid-Atlantic Ridge immediately south of Iceland. In the course of the survey, forty-nine parallel crossings had been made over the ridge crest, each separated by about eight kilometers. Vine showed profiles from four of these and the correlation was remarkable, as was the excellent fit with his computed models. This data had been collected at Heirtzler's suggestion, but ironically when Heirtzler and his French coworker Xavier Le Pichon had first published it in June 1966 they had used it to argue against the Vine-Matthews-Morley theory. The obvious symmetry of the magnetic anomalies about the ridge axis convinced Vine that Heirtzler and Le Pichon were wrong. But in characteristically modest fashion, he simply described the remarkable correlation as "encouraging."

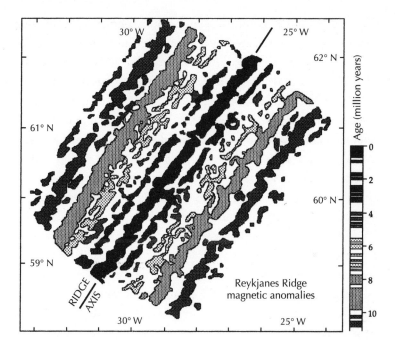

A chart compiled from a marine magnetic survey over the Reykjanes Ridge, part of the mid-Atlantic Ridge to the south of Iceland. Black and grey shading show areas where the magnetic field is stronger than expected, above normally magnetized seafloor. White shading represents areas of lower than expected field intensity above reversely magnetized seafloor. The whole pattern is symmetric about the ridge axis.

He went on to show equally remarkable fits to magnetic anomaly profiles observed over almost all the known spreading ridges: the South Atlantic, the Carlsberg in the northwest Indian Ocean, the Red Sea, an update of the Juan de Fuca, and the *Eltanin*-19 profile from the East Pacific Rise. He estimated spreading rates that varied from one centimeter a year for the Reykjanes Ridge and the Red Sea to 4.4 centimeters a year for the East Pacific Rise. A faster spreading rate meant that a polarity interval was represented by a wider strip of seafloor, and so even the shorter polarity events were more likely to be detected by a ship-borne magnetometer some 2000 or 3000 meters above the seafloor. In particular, the Juan de Fuca and East Pacific Rise profiles revealed considerably more detail than the others.

No longer lacking confidence, Vine now argued that the seafloor anomalies held the key to extending the geomagnetic polarity timescale: the process of seafloor spreading was laying out a complete history of Earth's magnetic field across the seafloor like a giant barcode in black and white, normal and reversed. Once you knew the spreading rate, you could read the code.

He tested this idea by assuming that his spreading rate of 4.4 centimeters a year was good for the whole 500 kilometers of the *Eltanin*-19 profile, and decoded the entire record. It turned out to be almost identical to the one Pitman and Heirtzler had published two weeks earlier. Apart from some slight revisions in dates, this timescale, covering the past eleven million years, is still accepted today.

Vine was not finished yet. He next pieced together magnetic anomaly profiles extending 3600 kilometers across the Pacific Ocean at a latitude of about 41° N, from the crest of the ridge at 127.5° W right out to 170° W, and estimated they represented a total of eighty million years of seafloor history, and an eighty-million-year record of changes in the polarity of Earth's magnetic field.

The stage was now set for the next big step. Who was going to take up the challenge? It would not be Vine. Working alongside Hess at Princeton, his interests were moving towards the geological implications of Wilson's plate tectonics. As chance had it, it would be Heirtzler—Pitman and Herron's mentor at the Lamont Geological Observatory—who stepped forward, but not before the seafloor had turned up yet another surprise to corroborate the evidence of the marine magnetic anomaly records.

As well as its potential length, the marine magnetic anomaly record had another advantage over lava flows dated through radioactive decay—continuity. With discrete, individually dated lava flows there was always the possibility of errors in ordering, and often gaps remained in the timescale where no lavas had been found. For example, the Gauss-Matuyama reversal was poorly dated for years, simply because no lavas of the right age had been studied. The seafloor timescale filled such gaps and also identified several short polarity "events" that had not been found in lava-flow studies.

The mid 1960s saw the launch of a coordinated international program of deep-sea drilling, and the first long cores of sediments from the seafloor began to come ashore. The process by which sediments become magnetized is quite different from the way magnetization is locked into volcanic rocks as they cool. With sediments, the tiny grains are already magnetized when they are washed into the ocean, and as they fall through the water they align, like a compass needle, with Earth's magnetic field. Eventually their orientation becomes locked by the surrounding grains as the sediment compacts.

Chris Harrison, a student at Cambridge, and Neil Opdyke, working along the corridor from Heirtzler at the Lamont Geological Observatory, were among the first paleomagnetists to study the records of magnetization in these cores. In 1966 Opdyke published results from seven cores of deep-sea sediment, each between four and twelve meters long, which had been collected around Antarctica. All were normally magnetized at the top, and reversely magnetized below a depth that varied from one to five meters. Some of the cores showed one or two normal events in this reversed interval, while all but one were predominantly normally magnetized again at their base. Marine microfossils, the remains of tiny bugs that lived on or near the seafloor, were used to align the cores with each other chronologically. As these bugs had evolved relatively rapidly, they provided a much better indication of age than the macrofossils traditionally used by hard-rock geologists to date older consolidated rocks.

When the cores were correlated on the basis of the microfossil changes, the correspondence of the polarity epochs and events jumped out immediately: despite differences in sedimentation rate, the cores carried the same continuous magnetic records through the Brunhes, Matuyama and part of the Gauss epochs. These sedimentary records provided valuable independent validation of the geomagnetic polarity timescale of Vine, Pitman and Heirtzler.

By now Heirtzler was an avid supporter of the Vine-Matthews-Morley hypothesis—and he was in an ideal position to extend the marine magnetic anomaly record to the very edges of the oceans, and the polarity timescale with it. Over the next two years he and his coworkers sailed the world collecting new long, high-quality magnetic anomaly records and decoded the history of the seafloor. In 1968 their results appeared in four papers published in a single issue of the *Journal of Geophysical Research*. Each of the new profiles was between 1500 and 3000 kilometers long and showed the now familiar sequences of wiggles, greatest closest to the spreading ridges, and gradually diminishing but remaining recognizable through their entire lengths.

The records showed that field reversals had been going on for at least the past eighty million years. To start with, Heirtzler and his team matched the first few anomalies of each profile to the first three-and-a-half million years of Doell and Dalrymple's polarity timescale—back to the Gilbert-Gauss transition. From this match they calculated a spreading rate for each profile. They then took a blind leap in the dark—the only real option open to them. On the assumption that each ridge system had spread at a constant rate throughout its entire history, they worked out the sequence of polarity reversals that best fitted each observed profile.

Astonishingly, the pattern of reversals—the black-and-white magnetic barcode—was the same in every case: the Pacific, Atlantic and Indian Ocean profiles yielded virtually identical patterns of normally and reversely magnetized seafloor. The reversal process was certainly global—only reversals of Gilbert's "magnus magnes" could cause simultaneous changes at all these sites.

Of course, there were some discrepancies. Extrapolation is a notoriously hazardous procedure, even when dealing with well-studied physical processes, and the mechanism of seafloor spreading was neither well studied nor well understood. There was little reason to suppose spreading rates were constant in the long term. Nonetheless, only five years after the publication of Cox's first polarity timescale—which consisted of three polarity intervals over the past three million years, and was based on just nine dated lava flows—Heirtzler and his coworkers had produced a continuous history of reversals going back eighty million years. This was still less than one-fiftieth of the age of the Earth, but there was a wealth of new information to challenge the theorists.

In the next forty years there would be big improvements in methods of dating rocks and sediments, and the accuracy and precision of the geomagnetic polarity timescale would undergo significant fine-tuning. Nevertheless, the basic features would remain unchanged.

First, the eighty-million-year-long "barcode" showed there had been 183 polarity reversals. In other words, Earth's magnetic poles had flipped, on average, every 430,000 years. Hospers" original estimate of half a million years had been remarkably close to the mark.

Second, the field's polarity had been surprisingly evenly divided between normal (49 percent of the time) and reversed (51 percent). However, the frequency of reversals had been anything but uniform: polarity intervals had ranged from 10,000 years to five

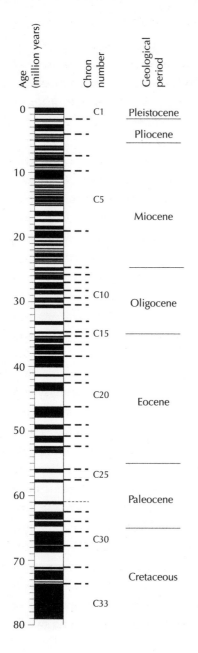

Age (million years)

Chron number

Geological period

0	C1	Pleistocene
		Pliocene
10	C5	
20		Miocene
30	C10	Oligocene
	C15	
40	C20	Eocene
50		
60	C25	Paleocene
70	C30	
80	C33	Cretaceous

■ normal polarity

▢ reversed polarity

A modern version of the geomagnetic polarity timescale for the past 80 million years, based on the 1995 work of Steve Cande and Dennis Kent at Columbia University's Lamont-Doherty Geological Observatory. The essential features are very similar to those of the original timescale that James Heirtzler and his colleagues deduced from seafloor magnetic anomalies and published in 1968. The conventional division into "chrons" and intervals of the geological timescale are shown to the right.

million, with most being less than 500,000 years. The median length was just 230,000 years.

There had been many more short intervals than long ones. Only twenty-one of the 183 intervals had lasted more than one million years; only twenty-five had been longer than the present Brunhes epoch. Paleomagnetists now suspect that much longer intervals of stable polarity, for which there is little or no seafloor data, occurred earlier in the planet's history. Other studies have shown, for example, that for most of the early Cretaceous period, from 83 to 118 million years ago, Earth's magnetic field was locked in a normal configuration, and it is suspected that during the Permian—225 to 280 million years ago—the polarity remained reversed for at least fifty million years, the so-called Kiaman Magnetic Interval.

It also seems that over a long timescale the frequency of reversals varies. Between forty and eighty million years ago there were only forty reversals—one per million years on average. However, in the past forty million years there have been 143 reversals, more than three times as many. This could be an artificial feature of the recording or preservation of the magnetization on the seafloor, or of the resolution of the measurements as the ship taking them moved away from the mid-ocean ridge into deeper water. However, geomagnetists don't think so: they believe it reflects some deep-seated change in the process by which the geomagnetic field is generated.

When it became clear that there had been a huge number of reversals, Cox's naming system based on names and locations, although nostalgic, seemed totally inadequate. In 1965, Walter Pitman had started to number the marine magnetic anomalies, and eventually the governing organization, the International Association for Geomagnetism and Aeronomy, recommended a new system based on an extension of this. The main polarity intervals were to be called "chrons." They would be numbered sequentially backwards from the present, and each would have a normal (N) and reversed (R) part. Thus, the Brunhes epoch became C1N, the upper part of the Matuyama C1R and so on. Shorter events, such as the Jaramillo, became "subchrons," and very long intervals of stable polarity, like the Kiaman, became "superchrons."

Was there any way of telling what happened during the process when Earth's magnetism reversed? Did the dipole simply tumble from one orientation to the other, or was the whole affair more complicated? The seafloor had provided few clues, other than that the process was too fast to be deciphered from the existing generation of marine records. Could the answer lie in deep-sea sediment cores, exposed sequences of sediments, or piles of lava flows? All were amenable to detailed sampling and the samples could be measured in the laboratory.

The first major finding from these studies was that the strength of the magnetic field dropped markedly—to about ten percent of its usual intensity—before its direction started to change. The declination and inclination then took about 5000 years to reverse before the field's strength recovered. The whole process seemed to take about 10,000 years.

It would prove harder to get consensus on *how* the direction changed. Because intensity was low in the transition periods, the magnetization, especially of sediments, was weak and difficult to measure precisely. Nevertheless, differences in the records of deep-sea cores coming from the various ocean basins soon showed that a simple tumbling dipole, however weak, was not the answer. When Earth's stable field weakened, the field that remained during the next five thousand years of a transition was a much more complicated one. As the theorists continued to grapple with ideas about the overall source of the field, these discoveries of the marine geophysicists and paleomagnetists had made their task even more complex.

In Search of a Solution

*The mechanisms behind the magnetic field and behind
the reversals are still really mysterious . . . one of the grand
intellectual challenges . . . in all of the physical sciences.*
—RAYMOND JEANLOZ, 1996

At the beginning of the twentieth century, despite hundreds of years of thought, speculation and experimentation, there was still no good idea of the origin of Earth's magnetic field. Even Albert Einstein reportedly declared it one of the greatest unsolved problems in physics.

In 1269 when Petrus Peregrinus had penned his *Epistola* and introduced the concept of the magnetic poles, he was not concerned whether the source of their "virtue" lay within the Earth, on its surface, or in the heavens: it was enough that the magnetic poles and celestial poles shared a common axis.

A little over three centuries later William Gilbert had declared the Earth to be *magnus magnes*—a great magnet—and so firmly rooted the origin of the magnetic field within the planet itself. He imagined that Earth carried a static, permanent magnetization "innate in all parts thereof and diffused throughout." Gauss had then used brilliant mathematical analysis to describe the main field in detail. He had clearly demonstrated that its source was internal, but had not probed much further into its origins.

If it were true that the Earth was uniformly magnetized throughout, it took only a simple calculation to show that the strength of that magnetization had to be some ten times that of the most strongly magnetized rocks found at the surface. By the nineteenth century, when William Hopkins and others were arguing about the make-up of the planet's interior, it was well known that temperature rose rapidly with depth. Below about twenty-five kilometers the temperature was, in fact, above the Curie temperatures of all known natural magnetic minerals, and so permanent magnetization must be all but impossible. But if only the outermost twenty-five kilometers of Earth were magnetized, the intensity of this magnetization would be unbelievable—greater than the strongest of today's super-magnets.

On top of this, secular variation, the gradual changes in Earth's magnetic field, had been discovered shortly after Gilbert's publication of *De Magnete*, and it was difficult to incorporate this into any theory based on permanent magnetization. Clearly then some other, dynamic source was at work.

To make things even more complicated, by the beginning of the twentieth century considerable effort had gone into studying the effects of the sun on a compass needle. It appeared the sun played some role in both the regular daily fluctuations—the tiny, rapid changes first noticed by London instrument-maker George Graham—and in the occasional magnetic storms that had intrigued Alexander von Humboldt in the early nineteenth century.

The confusion of ideas was aptly described in a 1903 textbook, *The Realm of Nature*, by H.R. Mill, a lecturer at Oxford University:

> Thus it appears that while the Earth's magnetism resides in the massive rocks of its crust, and is probably produced and maintained by the Earth's rotation, the sun's energy exercises a regulating or disturbing influence upon it.

The role of the sun, superimposed on the bigger picture of the main, internal field, was certainly a complication. However, it would be the discovery at the beginning of the twentieth century that the sun, too, had a magnetic field that would precipitate the next generation of ideas about Earth's magnetism.

Following James Clerk Maxwell's revelation that light was an electromagnetic wave, many physicists had studied the wavelengths or "lines" in the spectra of light emitted by the atoms of certain gases, and recognized the same wavelengths in the light radiated by the sun and other stars. A Dutch physicist, Pieter Zeeman, had shown that in a magnetic field a characteristic spectral line might be split into two or more lines of slightly different wavelengths. In 1908 George Ellery Hale, an American astronomer, would spot this "Zeeman splitting" in light emitted by hydrogen atoms in the atmosphere of the sun—particularly in light coming from sunspots which, being darker, were assumed to be cooler than the rest of the sun. It seemed that sunspots, as well as being cool, were also strongly magnetic.

Further studies would show that the sunspots occurred in pairs of opposite polarity: the magnetic field lines left the surface of the sun through one member of a pair and re-entered through the other.

In 1848 the German pharmacist and astronomer Samuel Heinrich Schwabe had discovered that sunspots varied in number in eleven-year cycles. It now seemed that the eleven-year sunspot cycle involved not just the number of the sunspots, but also their magnetic polarities. With this discovery, the magnetism of the sun suddenly became a hot topic of research. For a time it would almost eclipse the problem of Earth's magnetic field.

By now two sources of magnetism were known—permanent magnets, which had been recognized for at least two millennia, and electric currents, the magnetic properties of which had been discovered by Ørsted the previous century. But a new idea was emerging among atomic physicists: perhaps magnetism was intrinsically associated with rotation.

In 1912 Arthur Schuster, a physicist who had worked under Maxwell in the Cavendish Laboratory at Cambridge and later become professor of physics at Manchester University, reviewed all three as possible sources of Earth's magnetic field. Surprisingly, Schuster was reluctant to completely rule out permanent magnetism, arguing that although magnetization of the Earth's crust alone could not account for the observed field, it was not known what happened to Curie temperatures at the extreme conditions likely in the core, and so he kept open the possibility of more deep-seated permanent magnetization.

He was, however, quick to reject electric currents as a possible cause. For electric currents to exist in the Earth, he pointed out, they would have to be either maintained by permanent electromotive forces, or be the decaying remnant of currents that had existed since the Earth's formation. He could conceive of no driving mechanism for a permanent electromotive force, and argued that for the remnant of a primordial current to still exist the current must originally have been enormous. The origin of such a current, he said, did not bear examination, adding that whereas "we must keep our minds open to the possibility that the iron contained within the Earth is magnetizable . . . the difficulties which stand in the way of basing terrestrial magnetism on electric currents inside the Earth are insurmountable."

So what of the third option? Quantum physicists had recently discovered a connection between the spin, or angular momentum, of an elementary particle such as an electron, and its magnetic moment. The same was true for an atomic nucleus. From this had sprung the idea that magnetism might be intrinsically associated with rotation. Like the force of gravity, such magnetism might be difficult to detect on an everyday scale, but in bodies of astronomical proportions could it be significant? Schuster considered various ways this might come about but reached no firm conclusion.

The idea of rotation as the underlying cause of Earth's magnetic field languished. However, a new idea was about to burst on to the scene. Its instigator, Joseph Larmor, had been born in Northern Ireland and studied at Queen's University, Belfast and St. John's College, Cambridge before becoming Lucasian professor of mathematics at Cambridge in 1903. In his studies, Larmor was to forge a curious link between the classical physics of the nineteenth century and the quantum physics and relativity that sprang up at the beginning of the twentieth.

Outside academia Larmor displayed an equally enigmatic mix of progressive and conservative traits. In his maiden speech as Unionist member of parliament for Cambridge University, a seat he would hold from 1911 to 1922, he vehemently endorsed home rule for Ireland, yet in college politics he was distinctly retrograde, querying the need to install baths as the university had done without them for 400 years. (Nevertheless, once they were in place he took, without fail, a daily walk across the Cam to bathe in the New Court of St. Johns College.)

In 1919 Larmor was working on possible explanations for the magnetic fields of sunspots. One explanation, he suggested,

Joseph Larmor. While investigating the origin of magnetic fields associated with sunspots, this Irish-born professor of mathematics at Cambridge University came up with the idea that Earth's magnetic field might originate from a self-sustaining dynamo in the planet's liquid iron core.

might also relate to the Earth. He had found that when atoms were placed in a magnetic field, their magnetic moments started to revolve about the field direction, rather like the axis of a spinning top about the vertical. The discovery of this motion, now known as "Larmor precession," would eventually lead to the development of nuclear magnetic resonance, magnetic resonance imaging techniques and indeed the proton precession magnetometer. However, in 1919 Larmor's preferred explanation for the magnetic fields of sunspots was based not on rotation or precession, but on "an internal cyclic motion [acting] after the manner of the cycle of a self-exciting dynamo."

This was a flash of inspiration. Schuster had rejected the idea of electric currents in the Earth because he had been unable to account for their existence. Larmor, on the other hand, had thought back to Michael Faraday's law of electromagnetic induction and the principle of his disc dynamo. Faraday had shown that when a conductor moved through a magnetic field, a voltage was induced in it. If the conductor formed part of a closed circuit, this resulted in an electric current. If the magnetic field of that current reinforced the original magnetic field, the result was positive feedback and a self-exciting dynamo, which would, as Larmor wrote in a one-page note in *Reports of the British Association for the Advancement of Science,* "maintain a permanent magnetic field from insignificant beginnings."

However, nature does not give something for nothing; in fact, as a popular paraphrase of the second law of thermodynamics states, you can't even break even. As Peregrinus had found—although he did not recognize it as a law of nature and instead blamed his own ineptitude—perpetual motion simply does not occur. Without the continual injection of energy, the process will eventually run down.

Recognizing this problem, Larmor added the rider "at the expense of some of the energy of the internal circulation." At this point he was still talking about the magnetic fields associated with sunspots and relatively small-scale fluid circulation near the surface of the sun, but at the end of his note he commented that a similar concept would readily account for the "extraordinarily rapid" secular variation of Earth's magnetic field by "change of internal conducting channels," although it would require "fluidity and residual circulation in deep-seated regions."

He was clearly referring to the core of the Earth. Seismologists and geodesists had by now identified the boundary between the core and the mantle, and many were convinced the core was liquid.

Furthermore it was heavy, probably metallic, and therefore electrically conductive. As Larmor had realized, only a fluid could move rapidly enough to produce field changes typical of geomagnetic secular variation, features of which move across the globe at tens of kilometers per year.

Could such a self-exciting dynamo really work, either in the sun or in Earth's core? Over the next few years physicists grappled with the mathematics. This involved combining several theoretical constructs: Maxwell's equations, which dealt with both the induced electric currents and their associated magnetic fields; Ohm's Law, from which the energy required to maintain the currents would be determined; and equations describing the fluid dynamics of the core, including the so-called Navier-Stokes equations. These last equations stemmed back to the early nineteenth-century French mathematicians Siméon-Denis Poisson, Claude Navier and Augustin-Louis Cauchy, and to George Gabriel Stokes, another of William Hopkins' Cambridge senior wranglers.

Once the "magnetohydrodynamic" problem had been formulated mathematically, the task of solving it from first principles was daunting in the extreme. The sheer number of calculations required was so enormous that a full solution would not even be attempted for another seventy years. In the short term, progress could be achieved only by making simplifications—for example, by assuming a simple pattern of fluid flow and seeing whether the equations led to a self-sustaining magnetic field.

The pursuit of a dynamo solution almost stopped in its tracks in 1934, when Thomas Cowling, an English mathematician working on the sun's magnetic field, proved a dynamo could never result from fluid flowing in a symmetrical pattern about the rotation axis. This discouraging result became known as Cowling's anti-dynamo theorem.

In the face of this setback, it was not surprising that Schuster's third idea was revived—particularly when an American astronomer, Horace Babcock, announced that in addition to the local magnetic fields associated with sunspots, the sun also possessed an *overall* magnetic field, dipolar in form like the Earth's, and that he had also detected and measured magnetic fields in other stars.

Did this mean that magnetic fields were intrinsic to all massive rotating bodies? The chief proponent of this theory was the renowned professor of physics from Manchester University, Patrick Blackett, who was soon to be awarded a Nobel Prize for his pioneering work on cosmic rays. Blackett thought it was significant that the ratios of rough measurements of magnetic field strength (or, more precisely, dipole moment) and rotational angular momentum came very close to the square root of the gravitational constant—the constant, G, in Newton's law of gravitation, which had eventually been measured by Henry Cavendish—divided by the speed of light. He built up an elaborate theory around this. For a while this rotating body theory was dubbed the "fundamental theory" and received considerable attention.

Which theory was right? One test would be whether the strength of the magnetic field increased or decreased with depth into the Earth. According to the dynamo theory, the strength of the magnetic field should increase as its source in the core was approached. The rotating body theory suggested it should decrease. Keith Runcorn, then Blackett's research student, was assigned the task of measuring the magnetic field 1200 meters down the shaft of a disused Lancashire coal mine. The results, reported at a meeting of the Royal Astronomical Society on February 27, 1948, failed to answer the question. Blackett valiantly tried to keep the theory alive but Thomas Cowling, having already poured cold water on the dynamo theory, now criticized the rotating body theory for its "indefinite" basis.

Blackett was not defeated. He now attempted to prove the rotating body theory in a famous experiment designed to detect a magnetic field around a rapidly rotating gold sphere. (The sphere apparently had to be gold to make it heavy enough.) For this experiment he built an incredibly sensitive "astatic" magnetometer—essentially two identical magnets placed in opposite orientations and suspended one above the other on the same wire. Such an arrangement would detect the non-uniform magnetic fields that Blackett hoped to see around his sphere, while eliminating the effect of the prevailing laboratory field.

Unfortunately, although of unprecedented sensitivity, the magnetometer detected nothing, and Blackett finally had to concede victory to the dynamo theorists. His efforts had, however, not been entirely in vain: in the 1950s his magnetometer would become the nucleus from which British paleomagnetism flourished.

At the same 1948 meeting of the Royal Astronomical Society at which Runcorn had reported the results of his coal-shaft experiment, Edward Bullard, a post-doctoral researcher from Cambridge, had presented his new theory about the secular variation. First he described the regional nature of secular variation, with features growing, drifting and decaying in a matter of decades or centuries—much faster than most geological processes. Although these features were distinct from the main field, he reasoned they must also originate in Earth's fluid core.

He then demonstrated that if fluid near the core–mantle boundary circulated locally, a bit like an eddy, the presence of the main field would cause electromagnetic induction to occur in it, and this would generate local currents and magnetic fields just like those of the secular variation.

This was not a self-exciting dynamo: the main field was an essential ingredient to generate the secular variation. However, it was a major step forward for dynamo theory—that is, until the next speaker took the floor. Imperial College's John Bruckshaw proceeded to astound the audience by announcing that he and New Zealander Edwin Robertson had found reversely magnetized dykes in the north of England. Until this time the theorists had not taken the possibility of polarity reversals very seriously. If Bruckshaw were correct, the "geodynamo" problem had suddenly become much more challenging.

The Geodynamo

In this way it is possible for the internal cyclic motion to act after the manner of a self-exciting dynamo, and maintain a permanent magnetic field from insignificant beginnings, at the expense of some of the energy of the internal circulation.

—Joseph Larmor, 1919

The ball was now firmly back in the court of the dynamo theorists. The concept was gaining new momentum from the work of two physicists turned geophysicists. Walter Elsasser, born in Germany in 1904, had completed a PhD in the new discipline of quantum physics at the University of Göttingen, before emigrating to the United States and becoming an American citizen in 1940. After the war he took up a series of academic positions at prestigious universities across the country, but it was his geophysical work at the University of Pennsylvania between 1948 and 1958 that would earn him the reputation of father of geodynamo theory.

Edward (Teddy) Bullard, who had used electromagnetic induction to explain secular variation at the 1948 Royal Astronomical Society meeting, had studied nuclear physics under Ernest Rutherford at Cambridge before his interests turned to geophysics. After a year at the University of Toronto, he returned in 1950 to head Britain's National Physics Laboratory, where he became captivated by the geodynamo problem. He finally returned to Cambridge University in 1956 as head of the Department of Geodesy and Geophysics. He had missed the era of Hospers, Runcorn, Creer and Irving, but would be there in time for Vine and Matthews and the plate-tectonic revolution of the 1960s.

Elsasser and Bullard were both well acquainted with the magnetohydrodynamic equations and their complexity. They agreed that there must be a dynamo or regeneration mechanism, otherwise the field would die away in a matter of a few tens of thousands of years because of electrical resistance. Furthermore, they had both grappled with the problem of the origin of the energy that kept the regeneration process going. Having ruled out astronomical effects such as a difference between the moon's drag on the core and mantle, or precession or nutation of Earth's rotation axis, they concluded that only internally driven motion of the core fluid could be the answer. And this, they agreed, would occur only if there were a sufficient difference in temperature between the inner core and the base of the mantle.

With a complete solution of the equations still way beyond their capability, both independently opted to further investigate the mechanical analogue, the disc dynamo. It was easy to see that a simple modification of Faraday's disc dynamo would transform it into a self-exciting dynamo. Instead of using a separate magnet to provide the magnetic field, the field created by the output current was used; the apparatus merely had to be wired so the magnetic field was in the required direction—essentially a positive feedback effect. In principle, as long as energy was provided to keep the

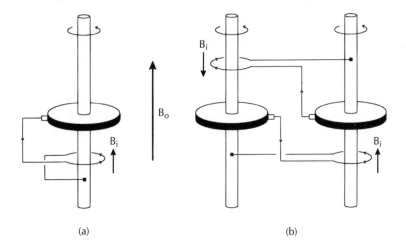

B_i

B_i

B_o

(a)

B_i

(b)

Single- and double-disc dynamos. In each case there is an initial
magnetic field, B_o, upwards. When each disc is rotated in this field,
a current is induced from its center to its rim. This is directed through
conducting wires that loop around the axles of the discs as shown.
In the single disc dynamo (a), the induced current produces a magnetic
field, B_i, that reinforces the initial field B_o, and so sustains the dynamo.
In the double-disc dynamo (b), the circuit is arranged so the current
induced in one disc provides the auxiliary magnetic field for the other,
but they are in opposite directions. In the right-hand disc B_i reinforces
B_o, but in the left-hand disc B_i opposes B_o—slowing it until eventually
the direction of rotation, and so the polarity of the dynamo, reverses.

disc rotating, such a system would continue to generate an electric
current and produce its own magnetic field. A small "seed" magnetic
field was needed to get the dynamo started, but then the seed could
be removed and the dynamo's own magnetic field would stabilize.

Unlike the geodynamo, the disc dynamo could be described
mathematically by relatively straightforward, solvable equations.
Hence, some of Bullard's early work drew analogies between his
solutions to the mathematics of the disc dynamo problem and
what might actually happen in the Earth's core.

Even so, as soon as Bullard tried to include the tendency of the magnetic field to decay with time, his equations became intractable. To get approximate solutions, he had to resort to a machine at Cambridge University called a differential analyzer. A mechanical forerunner of the electronic digital computer, this machine consisted of systems of rods and gear wheels that, once set, would solve differential equations at the crank of a handle. In this way Bullard was able to show that a disc dynamo produced a roughly dipolar magnetic field that oscillated in strength but maintained a single, stable polarity.

He readily acknowledged that beyond this point the analogy between the mathematics of the disc dynamo and of the geo-dynamo became "a little precarious." The rotor, disc and wires of a disc dynamo were a far cry from the Earth's core, with its large volume of mobile conducting fluid and solid inner part. In the language of the experts, the whole of the conducting core is "simply connected" and the geodynamo is therefore "homogeneous." Whereas in a disc dynamo (or a power station generator), the rotor is turned mechanically—by hand or by the action of water, wind or expanding pistons—in Earth's core the driving force had to be convection, the welling up of material from deep in the core towards the boundary between the core and the mantle.

Bullard's single-disc dynamo could produce a magnetic field in either direction, depending only on the direction of the initial seed field. Once the dynamo was going, the external seed field could be removed and the dynamo would continue to run happily. Oscillations in the strength of the current and the induced field occurred sporadically in some situations, but it turned out that the field's polarity remained stable: once up and running, the single-disc dynamo never underwent polarity reversals.

To achieve this extra trick, in 1957 Tsuneji Rikitake of the Earthquake Research Institute in Tokyo worked out the effect of

The differential analyzer at Cambridge University's Mathematics Laboratory, used by Edward Bullard to calculate solutions to the disc dynamo problem in the early 1950s. A mechanical forerunner of today's digital computers, the analyzer consisted of systems of rods and gear wheels, which, once set, solved differential equations at the turn of a handle.

a double-disc dynamo. Two discs rotated on parallel axes, but the coil carrying the current induced in the first was wound around the axis of the second and vice versa, so each disc provided the magnetic field for the other. In this way Rikitake tried to simulate the interaction that had to be taking place between different components of the magnetic field in the Earth's core.

Amazingly, this simple idea did just what Rikitake had hoped: of its own accord, the magnetic field flipped randomly from one polarity to the other—in what was possibly the first computer-generated example of a chaotic process. Was the double-disc dynamo really mimicking what went on in the Earth's core? Who knew, but it was an encouraging start.

Where would the dynamo scientists look next? Bullard, Elsasser and Rikitake had worked with mathematical models. While a fully realistic physical model of the core was clearly not possible, some progress could be made in laboratory experiments that simulated part, if not all, of the convection-driven, fluid dynamo.

In the 1960s Frank Lowes and Ian Wilkinson, working alongside Ken Creer in the research group that Keith Runcorn had now established at the University of Newcastle-upon-Tyne, laid to one side the fluid nature and convection of the geodynamo and focused their attention on the magnetic field. The model they built consisted of two cylinders made of an iron-rich alloy. These cylinders were free to rotate in a block of the same alloy, with which they were in good electrical contact via a thin layer of mercury. When the cylinders were made to rotate in a "seed" magnetic field, the currents and magnetic fields induced in each cylinder were found to sustain each other, in a manner similar to Rikitake's double-disc dynamo. However, there was a difference: Lowes and Wilkinson's model was a "homogeneous" self-exciting dynamo: there were no wires to channel the currents. A second version, in which the geometry was modified and minor changes made to such things as the materials and rotation speed, even went through the now essential spontaneous polarity reversals.

The next two decades saw the introduction and rapid increase in the use of electronic computers in many areas of research, but this boost in computational power served only to clarify the enormity of the magnetohydrodynamic problem. Geomagnetists were still forced to make assumptions and simplifications in order to run their calculations. One was the so-called "kinematic dynamo," in which the computer was given an initial pattern of fluid flow and programed to calculate the resultant magnetic field. However, the solutions could only ever be as good as the initial assumptions, and none was yet anywhere near Earthlike.

During this period, a group of Cambridge University students in white gloves were poring over library archives and ships' logs. Brought together by Dave Gubbins, their supervisor, their mission was to retrieve and collate every scrap of declination, inclination and intensity data ever recorded.

The students painstakingly pieced together the history of Earth's magnetic field as far back as 1590. Using these data they then calculated the field at the boundary of the mantle and the core. The results showed that the magnetic field at the core–mantle boundary was much more complex than the field observed at the Earth's surface. Amid the general complexity, one feature stood out: much as Halley and Hansteen had observed earlier for the surface field, there appeared to be not two but four locations at which the magnetic field lines were concentrated. Two were in the northern hemisphere, where the field lines entered the core, and two in the southern hemisphere, where the field was outwards. None of these locations lay on Earth's rotation axis, which indicated there was a complicated pattern of convective motion within the outer core. This was another feature against which geodynamo models had to be tested.

In 1986 Englishman Paul Roberts and American Gary Glatzmaier met and began working together on the geodynamo problem. Roberts had recently joined the University of California at Los Angeles (UCLA) as professor of mathematics and geophysical sciences. He would later describe himself as "a doyen of geomagnetic theory, one of the old fogies of the subject:" his work in geomagnetism went back to the days of Elsasser and Bullard.

At the outset of Roberts" graduate studies at Cambridge University in the early 1950s, his supervisor had proposed he should "either prove that fluid dynamos could not exist, or find a working model." Bamboozled by the magnitude of the demand, he had promptly changed his thesis topic to the slightly more tractable problem of the origin of the secular variation and found a new supervisor—none other than Keith Runcorn, who was presumably delighted to take on yet another such able student.

Roberts' enthusiasm for the geodynamo problem as a whole had been reignited in the 1960s, by which time he was working in the geomagnetism group at Newcastle upon Tyne and news was leaking out from behind the Iron Curtain that a Soviet geophysicist, Stanislav Braginsky, had cracked some of the long-standing issues and discovered a set of almost symmetrical solutions. Meantime, new results on the magnetic fields of the sun and other stars were accumulating thick and fast, and it was now known that Jupiter too had a strong dipolar field, and presumably an internal dynamo. Roberts began to work tirelessly to advance the theory of kinematic dynamos, to understand the balance between the various forces acting on the core fluid—iron-rich metal in the case of the Earth, but "metallic" hydrogen in the case of the sun and Jupiter. Once again, he was drawn towards the grail of deriving a fully self-consistent homogeneous magnetohydrodynamic model of Earth's core—with no prior assumptions regarding fluid flow, and minimal simplifications.

When Roberts arrived at UCLA, Gary Glatzmaier had just finished writing his PhD thesis—on convection in a rotating spherical shell —at the University of Colorado, and had joined the United States National Laboratory at Los Alamos in New Mexico, where he was engaged in modeling magnetohydrodynamic processes in the sun. Computers had come a very long way since Bullard had first used the mechanical differential analyzer to solve his differential equations and since Vine and Matthews had fed their seafloor models on paper tape into one of its early electronic successors, the Edsac 2.

In 1995, after several years of collaboration, Glatzmaier and Roberts seized the opportunity to harness the unprecedented speed and capacity of the new Cray C90 supercomputer at Pittsburgh University. Their goal was to produce a computer simulation of Earth's magnetic field that went through all the processes and reproduced all the features that had been documented—not just from the four hundred years of direct, first-hand observation, but from the paleomagnetic records that stretched back to the beginning of the planet's history. Their results would stun the world of geophysics.

To make the computer mimic all the complex processes taking place in Earth's core, Glatzmaier and Roberts programed it to calculate temperature, pressure and density, the velocity of the fluid, and the magnetic field by simultaneously solving all the equations of the magnetohydrodynamic problem. Using spherical geometry—latitude, longitude and radius—the computational problem was similar in principle to, but enormously larger than, the one that Gauss's team of "calculators" had taken on in the 1830s.

A single solution, a snapshot for one instant of time, would not be sufficient. Since the geodynamo evolves with time, the equations had to be solved over and over, stepping forward by the equivalent of twenty days each time. And as the solutions corresponding to different times depended on each other, it was necessary to look backwards and forwards at each step of the calculation. It was this more than anything else that took up so much computer time: the exercise would have been all but impossible before the advent of the supercomputer.

Finally, the solutions had to be converted to values of temperature, pressure, density, fluid velocity and magnetic fields at locations on a huge spatial grid: at a total of 100,352 points in the outer core (49 different radii at each of 32 latitudes and 64 longitudes) and 34,816 points in the inner core (17 radii x 32 latitudes x 64 longitudes).

Initially, Glatzmaier and Roberts ran their program for the equivalent of 40,000 years. At twenty-day intervals, this amounted to 730,000 steps, and a total of 100,000 million calculations for each component. The process took over a year, and required 2000 hours of central processor time on the Pittsburgh supercomputer.

The results were incredible. Graphic representations of Earth's magnetic field, such as the one reproduced on page 234, would become icons of late twentieth-century science and grace the covers of numerous books and journals, including *Science* and *Nature*.

After the equivalent of about 10,000 years, the model settled so that the field at the Earth's surface was predominantly dipolar. So far, so good. However, the image of the field emerging from the core–mantle boundary showed the surface field was just a smoothed version of a much more complicated picture.

At the core–mantle boundary, bundles of magnetic field lines emerge at high latitudes, but just as Gubbins' students had found in their extrapolation of historical data to this boundary, there are two bundles in each hemisphere, and they are not actually at the poles. In fact, at the poles of the core–mantle boundary the field is relatively weak.

Within the core, the picture is even more complicated. As well as the field lines emerging from the core, which make up what is known as the "poloidal" field, others are wrapped up inside the core like tangled spaghetti. This "toroidal" field is locked within the core, and since its field lines never penetrate the core–mantle boundary, it is impossible to detect from outside. As time progresses, the twisting and shearing caused by convection and the rotation of the Earth convert toroidal field lines into poloidal field lines and vice versa—essential processes in the operation of the geodynamo.

A snapshot of the magnetic field lines inside and near Earth's surface during a period of stable, normal polarity, as modeled by Glatzmaier and Roberts. The inner circle represents the core–mantle boundary, and the outer circle the surface of Earth. White and grey lines represent inward-directed and outward-directed magnetic field lines respectively. Outside the core the familiar dipole field predominates, with field lines exiting the core and Earth's surface in the southern hemisphere, and looping round to re-enter the surface and core in the northern hemisphere. Within the core, the additional "toroidal" magnetic field makes the pattern much more complicated.

Gary Glatzmaier (left) and Paul Roberts. The two men began working together in 1986 on the problem of how to create an accurate model of the processes taking place in Earth's core. In 1995 they seized the opportunity to use a new supercomputer at Pittsburgh University to carry out calculations that would previously have been impossible, and ultimately solved the mystery of what causes Earth's magnetic field.

At the Earth's surface, the simulated magnetic field goes through changes that look very much like observed secular variation, with features growing and decaying, while generally drifting westward at a rate close to the 0.2 degrees a year that Bullard measured in the early 1950s. This, however, belies what is going on in the outer core, where the field is quite unstable, and in fact regularly tries to reverse polarity. Most attempts to reverse polarity are unsuccessful only because the inner-core fails to follow suit: it has a longer magnetic decay time, and so takes longer to respond to the driving field from the outer core. More often than not, the outer core "changes

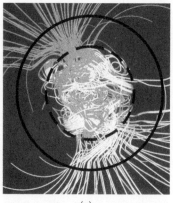

(a)

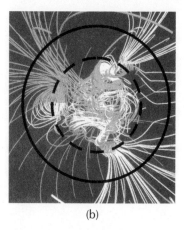

(b)

(c)

Glatzmaier and Roberts' simulation of Earth's magnetic field during a polarity reversal from reversed to normal. In each case, the inner circle represents the core–mantle boundary and the outer circle the surface of Earth. In (a) the field is still predominantly dipolar, but already tilted significantly from the geographic rotation axis. In (b) the pattern of field lines is more complicated, and the field intensity at Earth's surface has weakened considerably. In (c) the dipole is beginning to regenerate in the opposite (normal) direction, although a complex field structure persists, particularly in the southern hemisphere.

its mind" and flips back before the inner-core field can react, and so a polarity reversal does not occur. In this way, the electrically conducting properties of the inner core has a crucial stabilizing effect on the geomagnetic field as a whole.

Nonetheless, it was right at the end of its run that Glatzmaier and Roberts' computer simulation produced its most startling event. After the equivalent of nearly 40,000 years, the outer-core field destabilized and waited just long enough for the inner-core field to reverse too. Instead of regenerating in the original direction, the whole magnetic field flipped and grew back in the opposite polarity. The computer model had undergone a full geomagnetic polarity reversal.

Glatzmaier and Roberts' dream had come true. The simulation had maintained a stable dipole field for 40,000 years. It had mimicked secular variation in every required manner. It seemed to have tried to reverse several times but had almost always reverted to its original polarity. And eventually, just once, it had succeeded in executing a full polarity reversal.

The two men had answered Einstein's ninety-year-old challenge. In landmark papers in the journals *Nature* and *Physics of the Earth and Planetary Interiors*, they convinced the world of science that electric currents in the Earth's molten, iron-rich outer core, brought about by the combined churning effects of convection and the planet's rotation, were all that was needed to account for everything known about Earth's magnetic field—that and a very large and powerful computer.

Epilogue

The Earth is not simple . . . Significant breakthroughs . . .
will come from young people who are inspired to work
on what others proclaim is impossible.

—GARY GLATZMAIER, 2002

Amazing as Gary Glatzmaier and Paul Roberts' achievement was, it would mark not an end but the beginning of a new era of Earth science, one in which the geodynamo was no longer mere theory. Their computer model had demonstrated that a magnetohydrodynamic dynamo that obeyed Maxwell's equations of electromagnetism and Navier and Stokes' equations of fluid dynamics could generate an Earthlike magnetic field. Their 1996 papers unleashed a whole new burst of enthusiasm and activity.

First, the world watched closely as Glatzmaier and Roberts' model chugged through more and more computer time, but after the equivalent of 300,000 years it had still not achieved another polarity reversal. This was not altogether unexpected: in real time it has been 780,000 years since the last documented reversal, and if paleomagnetism has told us anything it is that the past is no indication of the future. Polarity reversals occur randomly, and the intervals between them range from a few thousand to tens of millions of years. Perhaps Glatzmaier and Roberts had been lucky the first time round.

But what about the repeated attempts of their model's outer-core field to reverse that had been thwarted by the failure of the inner-core to follow suit? Examination of paleomagnetic records has revealed there have been several major "excursions" from the stable magnetic field direction, much bigger than regular secular variation but not always recorded worldwide, and not eventuating in full polarity reversals. Examples are the so-called Laschamp excursion, recorded in 40,000-year-old lava flows in the Chaîne des Puys region of France, and the Mono Lake excursion, thought to have occurred in California about 25,000 years ago. Although replicated and credited by many as genuine features of the geomagnetic field, neither the Laschamp nor Mono Lake excursions had been found in paleomagnetic records from sites some distance away. The significance of these regional excursions had proved elusive, but now here was an explanation: attempts of the outer core to reverse polarity that had been aborted because the inner core had failed to follow suit.

The picture was, however, still far from complete; many questions remained unanswered. Glatzmaier and Roberts' model had been successful, but the limitations of computing power had forced them to assign unrealistic values to some of the quantities. Other researchers now began to publish their own geodynamo models and gradually close the gap between computation and reality, but as Glatzmaier wrote in the 2002 *Annual Review of Earth and Planetary Science*:

> Recent geodynamo simulations have provided new insights and predictions for convection and magnetic field generation in the Earth's core . . . [but] . . . we still have a long way to go.

The computational costs of producing a fully realistic geodynamo model would, he pointed out, be staggering by today's standards.

A major question remained about the source of the energy required to drive the dynamo—the mechanism behind convective motion in the outer core—and how it should be modeled. Compared to convection in the Earth's virtually solid mantle, which is a sluggish affair resulting in seafloor spreading rates of only a few centimeters a year, convection in the outer core, where the fluid is thought to flow as easily as water, is up to a million times faster—millimeters per second, or tens of kilometers per year. What drives this motion?

There is one obvious source: the heat imparted to the core when the Earth first formed. A large enough temperature difference between the inner core and the core–mantle boundary would drive thermal convection. Heat supplied by decay of radioactive elements in the core would aid the process, but it is uncertain whether a sufficient source of radiogenic heat exists within the core.

In addition to this, there is a process known as "compositional convection." The core is not only much denser than the mantle, it is also made of vastly different material. This, together with the fact that pressure increases enormously towards the center of the planet, has led geophysicists to believe that as the Earth has cooled from its original completely molten state, the inner core has gradually formed by "freezing" from the center outwards. The process is ongoing: the inner core presumably continues to grow as the Earth cools further and outer-core fluid progressively solidifies at the boundary between the outer and inner cores.

The inner core is also known to be denser than the outer core, and as outer-core material solidifies on to the inner core, lighter, albeit minor, components of the core fluid such as sulfur and oxygen are released. Just as a cork held under the surface of water will rise upwards when released, this lighter material, being buoyant, will rise naturally in the outer core—the movement known as compositional convection. Solidification also involves the release of heat (it is the opposite of melting, which requires heat input) so thermal convection will be further enhanced by this process.

Although scientists now agree that both thermal and compositional convection contribute energy to the geodynamo, the balance between these two processes, and how to incorporate them in models, is still intensely debated.

Another unsolved mystery is what actually happens during a polarity reversal. Computer simulations and paleomagnetic records agree that the field intensity is reduced considerably. Does this mean that Earth's cosmic ray shield is weakened, to the extent that it can no longer defend life on Earth from space's bombardment of charged particles and radiation?

Some scientists believe this is the case, and is what has led to genetic mutations and extinctions of whole species. Is there evidence to correlate geomagnetic polarity reversals with such extinctions? Certainly not on a simple one-for-one basis. Mass extinctions—for example, the demise of the dinosaurs 65 million years ago, at the famous Cretaceous–Tertiary boundary—have not occurred close enough to reversals that we can validly claim cause and effect. Given the complex interrelations we know must exist between the various Earth processes, it is probably naïve to expect such a simplistic correlation.

In the computer models, polarity reversals occur at seemingly random intervals of time. This is compatible with statistical analyses of the geomagnetic polarity timescale, and is the outcome of what mathematicians call a chaotic process. Like Gauss, Glatzmaier and Roberts were content to let the computations tell the story. This is not the case with most of us: we need a physical picture on which to pin mathematical models, so we insist on searching for a physical trigger that will initiate the necessary changes in core fluid motions and make the field try to reverse.

One candidate is what Richard Muller, a physicist at the University of California, Berkeley, describes as an avalanche on the core–mantle boundary. Muller has argued that if a massive glob of the slushy material on the boundary were suddenly to slip back into the core it would seriously disrupt the pattern of convective flow, possibly enough to initiate a reversal. Such an avalanche, Muller suggests, could be triggered by the jolt of an astronomical impact—for example, a large asteroid hitting the Earth.

This sounds familiar. The theory that a massive asteroid impact preceded the Cretaceous–Tertiary boundary extinctions is well known. The 200-kilometer-in-diameter Chicxulub crater, buried beneath the Yucatán Peninsula and Gulf of Mexico, is thought to be the site of an impact that threw enough dust into the atmosphere to cause a "nuclear winter" and kill off the dinosaurs. Muller's hypothesis of a chain of events from impacts to core–mantle boundary avalanches, mass extinctions and geomagnetic reversals is seductive in its logic. But is it backed up by observation, theoretical calculation and statistical correlation? Only time and hard work will tell.

Although from our perspective Earth appears unique, magnetic fields are now recognized as a common feature of many planets. In the 1950s, Jupiter's strong field was discovered by radio astronomers. In the 1980s, magnetometers carried by the Voyager spacecraft detected similar fields around Saturn, Uranus and

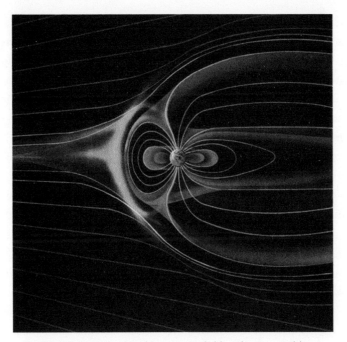

An artist's impression of Earth's magnetic field as if it were visible from space. On the left, the solar wind—a stream of charged particles emanating from the sun—is safely deflected around the planet by the magnetic field. As a result the field is compressed on Earth's day side, and drawn out into a long tail on the night side. When the solar wind is particularly strong, charged particles sometimes enter the upper atmosphere along field lines near the poles and produce aurora— spectacular light displays.

Neptune. The sun, Jupiter and Saturn are all thought to have dynamos based not on an iron-rich core like Earth's, but on hydrogen in a metallic state. The fields of Uranus and Neptune are subtly different and their explanation is not as simple. Few scientists doubt the existence of dynamos, but neither molten iron nor metallic hydrogen seems likely. One suggestion is that water at a high enough pressure may have the necessary properties.

Mercury is another intriguing case. In the mid 1970s, the Mariner spacecraft flew by the planet and picked up a weak magnetic field. Like Earth, Mercury is now thought to have an iron-rich liquid outer core, and so may conceivably sustain a similar dynamo. In 2004, NASA dispatched the *Messenger* spacecraft (named for *Me*rcury *S*urface, *S*pace *En*vironment, *Ge*ochemistry and *R*anging) to further investigate Mercury's magnetism. It has already flown by the planet a number of times, adjusting its path in the process, and will go into orbit in March 2011.

Venus has posed some curious questions. Despite this planet's similarity to Earth in size, mass and internal composition, and the near certainty that it has a suitable outer core, all attempts to detect a magnetic field have drawn a blank. Dynamo experts infer that for some reason the all-important convective motion is not happening in Venus's core. Some suggest Venus may not have an inner core.

However, perhaps the most interesting bodies in the inner solar system are the moon and Mars. On both, surface rocks seem to be strongly and permanently magnetized. Portions of the Martian surface even seem to display a magnetic barcode—alternating stripes of oppositely magnetized rocks reminiscent of Earth's seafloor. Mars no longer has an internal magnetic dynamo, but it may have had one in the past. Perhaps an initially molten core cooled and finally froze solid, switching off the Martian magnetic field in the process. If so, did Mars once experience both polarity reversals and plate tectonics? Perhaps Mars is more Earthlike than we have realized.

What about the planets outside our solar system, which are being discovered by astronomers at an increasing rate? The field of planetary dynamo theory is wide open. As New Zealander David Stevenson, professor of planetary science at the California Institute of Technology, has said, the whole issue is inextricable from "the big questions of how planets form, differentiate and evolve."

After centuries of scientific endeavor, it is clear that the planet on which we live is a complex and dynamic body. Mechanical, thermal and electromagnetic interactions occur ceaselessly between the inner core, the outer core, the mantle and the crust. Modern technology such as satellites and computer simulations allow us an ever greater understanding of these processes, and of how they create the unique magnetic field without which life on Earth could not exist. However, direct proof of the geodynamo remains the realm of science fiction. We will almost certainly colonize other planets before making Jules Verne's fantastic journey to the center of the Earth and learning the whole truth.

And what of the near future? In the past 200 years the strength of Earth's magnetic field has dropped by some fifteen percent. This rate of change has astonished many scientists. If it is sustained, they argue, we could right now be headed for the next polarity reversal when magnetic north becomes south and south becomes north.

Sometimes I let my imagination fast-forward: what if we were here to see this happen? Countless migratory species—birds, butterflies, whales, fish, even honeybees—use Earth's magnetic field lines as invisible way-finders on their great journeys across the planet. In the weakened transitional field, would these creatures increasingly lose their way and end up wandering the world like confused drunkards? If the magnetosphere surrounding the Earth were to collapse, would we humans be entertained by spectacular aurora at equatorial latitudes? Or would we instead be forced to take refuge from the onslaught of the solar wind?

No one really knows. What we do know is that life on Earth has survived hundreds of reversals of the magnetic poles up to now, and so the next is unlikely to kill us off completely. Someone, although not you or me, will be around to see the field rebuild and the compass needle swing to the south. This time, though, they will understand why.

Glossary

amber Fossilized tree resin. When rubbed with fur, amber acquires a negative electric charge, enabling it to attract light objects.

antiferromagnetism Form of magnetism in which the atomic magnetic moments of a material are ordered so that neighboring moments oppose one another and cancel each other out; the material then has no overall magnetization.

argon–argon dating method Modification to the potassium–argon method: a radiometric technique for estimating the age of a rock.

aurora Curtainlike display of colored light occasionally observed in the night sky, most often at high latitudes. Auroras are caused by charged particles emitted from the sun, and guided by Earth's magnetic field lines into polar latitudes, where they collide with atoms and molecules of the upper atmosphere.

baked contact Layer of rock heated by contact with hot volcanic lava, which often results in remagnetization in the direction of the magnetic field at the time of the lava's cooling.

Chandler Wobble Very small oscillation of Earth's axis of rotation with respect to Earth's surface, caused by the planet not being perfectly spherical.

charge (electric) Intrinsic physical property of some fundamental particles: the electron carries a negative charge (-1.6×10^{-19} C); the proton carries a positive charge ($+1.6 \times 10^{-19}$ C).

Coulomb's Law *See* **electrostatic force** and **magnetostatic force**.

compass Pivoted or suspended magnet or magnetized needle, balanced to swing in a horizontal plane so that it settles with its north pole pointing towards magnetic north.

continental drift Hypothesis that the present arrangement of continents resulted from the break-up of a large land mass and subsequent drift of the fragments over the surface of the Earth. Continental drift is explained by the theory of plate tectonics.

convection Fluid motion driven by density differences and gravity: less dense material in a fluid rises relative to more dense material. Thermal convection arises because of the tendency of hot material to expand, so becoming less dense and rising through cooler, more dense material. Convective motion that is not necessarily related to temperature differences within the fluid is known as compositional convection.

crust Rigid, rocky outermost layer of the Earth. Beneath the continents the crust is thirty to forty kilometers thick, but the oceanic crust is only about ten kilometers thick.

Curie temperature Characteristic temperature above which a ferro- or ferrimagnetic material loses its spontaneous (remanent) magnetization and becomes paramagnetic.

current (electric) (Rate of) flow of electric charge.

declination Angle between magnetic north and true north at any given location; equal to the deviation of the compass from true or geographic north (originally called variation).

diamagnetism Magnetic property whereby, when an external magnetic field is applied to a substance, a very weak magnetization is induced in it in the opposite direction to the external field; the magnetization is lost when the inducing field is removed.

dip Old term used for inclination.

dip needle Magnetized needle, pivoted at its center of mass so it can rotate about a horizontal axis; when aligned in a (magnetic) north–south plane, the needle comes to rest inclined to the horizontal at an angle equal to the inclination.

dipole *See* **electric dipole** and **magnetic dipole**.

dipole field (1) Magnetic (or electric) field due to a magnetic (or electric) dipole. It can be visualized as a series of magnetic (or electric) field lines looping from the north pole (positive charge) to the south pole (negative charge). (2) Part of Earth's magnetic field that can be modeled by a (tilted) geocentric dipole.

disc dynamo Mechanical dynamo consisting of a metal disc rotating in a magnetic field; a current is induced between the center and the rim of the disc and may be fed into a circuit. In a self-sustaining disc dynamo, the induced current is used to generate a magnetic field that reinforces the original magnetic field.

diurnal (daily) variation Variations in the direction and intensity of Earth's magnetic field, with a period of twenty-four hours; due mainly to heating of the atmosphere by the sun.

dynamo Original term for an electromagnetic generator: a device that converts mechanical energy into electrical energy through the process of electromagnetic induction.

effluvium/effluvia Imaginary fluid/s once thought to be responsible for the transmission of electric and magnetic forces.

electric dipole Positive and negative charges of equal magnitude separated by a small distance.

electric field In general, a region of space in which an electric effect can be observed; technically, the electric field at any point in space is the electrostatic force acting on a unit positive charge at that point.

electric motor Device that converts electrical energy into mechanical energy—for example, to turn the rotor of a machine.

electromagnetic induction Process of inducing an electric current by moving a magnet close to a conducting circuit, or changing the magnetic field through it.

electromagnetic wave Wave consisting of magnetic and electric fields that oscillate perpendicular to each other and to the direction of propagation. The spectrum of electromagnetic waves ranges from radio waves (low frequency, long wavelength) to gamma rays (high frequency, short wavelength); all electromagnetic waves travel at the speed of light, about 300 million meters per second.

electrostatic force Force that exists between two electric charges: it is proportional to the product of the charges, and inversely proportional to the square of the distance between them. Like charges repel, opposite charges attract (Coulomb's Law).

equator, geomagnetic The great circle on the surface of the Earth that lies 90 degrees from both the geomagnetic north and south poles.

excursion, geomagnetic Major change in the direction of Earth's magnetic field, usually lasting only a few thousands of years and often observed only regionally, after which the field returns to its previous stable polarity.

ferrimagnetism Spontaneous magnetic property of some natural minerals such as magnetite and titanomagnetite, certain alloys and ferrites, which may result in a strong stable remanent magnetization, with a characteristic Curie temperature. In ferrimagnetic minerals the atomic magnetic moments are ordered in two opposing directions, as in an antiferromagnet, but in unequal proportions so there is a residual magnetization.

ferromagnetism Magnetic property occurring in metals such as iron, cobalt and nickel, due to the spontaneous alignment of atomic magnetic moments, resulting in a strong, stable remanent magnetization. Above the Curie temperature, the spontaneous alignment is lost and the material becomes paramagnetic.

geocentric axial dipole *See* **geocentric axial dipole field**.

geocentric axial dipole field Magnetic field that would result from a dipole at the center of the Earth, aligned with the rotation axis. The declination of such a field would be zero at all locations on the Earth; both inclination and intensity would depend on latitude.

geographic poles The two locations where Earth's rotation axis intersects the Earth's surface; all lines of longitude (meridians) converge at the geographic poles.

geomagnetic field reversal Reversal in orientation of Earth's magnetic field, such that north and south geomagnetic poles change positions. *Also known as* polarity reversal.

geomagnetic polarity timescale (GPTS) Dated sequence of geomagnetic polarity reversals, derived mainly from marine magnetic anomaly profiles, radiometrically dated rocks and sedimentary sequences.

geomagnetic poles The locations where the axis of the geocentric dipole that best fits Earth's magnetic field intersects the Earth's surface; the north and south geomagnetic poles are antipodal.

gravitational force Force of attraction between two bodies that results solely from their masses. It is proportional to the product of the masses of the bodies, and inversely proportional to the square of the distance between them (Newton's Law).

Halleyan line Original name for an isogonic contour, or (imaginary) line of constant declination on the Earth's surface.

hematite A naturally occurring antiferromagnetic form of ferric oxide Fe_2O_3, mined as the main ore of iron.

homogeneous dynamo Dynamo process involving currents in a continuous medium, rather than in wires of a circuit.

inclination Angle of direction of the magnetic field below the horizontal; originally called dip. Equal to the angle of a dip needle below the horizontal.

induced magnetization Magnetization acquired by a material in the presence of an applied magnetic field, which is lost when the material is removed from the inducing field. This is the only form of magnetization found in paramagnetic and/or diamagnetic minerals. *See also* **remanent magnetization**.

inner core Innermost part of the Earth; a solid sphere about 1200 kilometers in radius, thought to be predominantly iron-nickel alloy.

International Geomagnetic Reference Field (IGRF) Mathematical model of Earth's geomagnetic field, based on spherical harmonic analysis; endorsed by the International Association for Geomagnetism and Aeronomy and updated every five years.

inverse square law Any law of physics in which a property decreases in proportion to the inverse square of distance. For example, if the distance is doubled, the property decreases to one-quarter of its original size; if the distance is tripled, the property goes down by a factor of nine; and so on. The force of gravity between two bodies, the electrostatic force between two charges and the magnetic force between two poles are all inverse square laws: each depends on the inverse square of the separation —of the bodies, charges or poles.

latitude Angle, measured at the center of the Earth, between a location on the surface and one on the same meridian (line of longitude) at the equator.

line of force A line drawn in a magnetic or electric field so that at every point it gives the direction of the field at that point.

lodestone Naturally occurring, strongly magnetized rock, rich in magnetite or titanomagnetite.

longitude The angle, measured at the center of the Earth, between the point where the meridian through a location crosses the equator and where the prime meridian crosses the equator.

magnetic anomaly Difference between the measured magnetic field at a location and that expected from the main geomagnetic field. The difference usually arises from remanent or induced magnetization in local rocks.

magnetic dipole In theory, magnetic north and south poles of equal strength, separated by a small distance. In practice, a bar magnet, a uniformly magnetized sphere or terrella, and a circular coil carrying a steady current each produce a dipole field.

magnetic equator Imaginary line on the surface of the Earth along which the inclination is zero.

magnetic field In general, a region of space in which a magnetic effect can be observed. Technically, the magnetic field at any point is related to the force experienced by a magnetic pole or a moving electric charge at that point.

magnetic moment Measure of the strength of a magnetic dipole.

magnetic poles (of Earth) The points on Earth's surface at which the magnetic field is vertical—that is, the inclination is +90° at the north magnetic pole and –90° at the south magnetic pole. The magnetic poles are, in general, not antipodal.

magnetic storm Irregular small-scale variations of Earth's magnetic field, particularly at high latitudes; associated with solar activity and often synchronous with auroras.

magnetite Naturally occurring mineral, an oxide of iron, Fe_3O_4, possessing strong ferrimagnetic properties.

magnetohydrodynamics Study of the movement of electrically conductive fluids in magnetic fields; the dynamics are determined by Maxwell's equations and the Navier-Stokes equations.

magnetosphere Volume surrounding the Earth inside which the geomagnetic field is confined by the pressure of the solar wind.

magnetostatic force The force between two static magnetic poles. It is proportional to the product of the pole strengths, and inversely proportional to the square of the distance between them; like poles repel, opposite poles attract (Coulomb's Law).

mantle The layer between the Earth's outer core and the crust; about 2900 kilometers thick, it makes up more than 70 percent of the Earth's volume. The mantle is almost solid and largely composed of silicate minerals.

mantle convection Very slow movement of Earth's mantle material, due primarily to upward heat flow from the core–mantle boundary and heat loss through the crust. The mantle is extremely viscous (almost solid), and convects very slowly in response to the temperature gradient. Hot buoyant material wells up at mid-ocean ridges, and cooler denser material is drawn back into the mantle along subduction zones.

Maxwell's equations Set of four mathematical equations describing the interactions between electric and magnetic fields, electric charge and current. Formulated by James Clerk Maxwell, they encapsulate the work of Faraday, Ampère and Gauss, and are the cornerstone of electromagnetism.

meridians (geomagnetic) Great circles passing through both geomagnetic poles.

meridians (geographic) Great circles passing through both geographic poles; lines of longitude.

mid-ocean ridge/rise Underwater mountain range rising several thousand meters above the seafloor, where upwelling mantle material forms new seafloor, which is then carried away in both directions by the process of plate tectonics.

natural remanent magnetization In situ remanent magnetization of a natural material, for example a rock.

Navier-Stokes equations Mathematical equation used to describe the dynamics of fluid flow.

non-dipole field (of Earth) Part of the geomagnetic field remaining when the field of the best-fitting tilted geocentric dipole is subtracted from the total field.

nutation Small nodding-like oscillation of Earth's rotation axis.

ocean trench Trench or valley that occurs on the ocean floor at a plate boundary, where one plate is drawn down beneath the other. *Also known as* **subduction zone**.

Ohm's Law Physical law which states that the electric current flowing through a conductor varies in proportion to the electric potential difference or voltage across it. The ratio of potential difference to current is called the electrical resistance, and is measured in ohms.

outer core Outer part of the Earth's core, between 2900 and 4100 kilometers below the surface. The outer core has a molten metallic composition, mainly iron, nickel and small amounts of various lighter materials.

***P* wave** Longitudinal seismic wave, often referred to as the **primary wave** because it is the first wave felt after an earthquake. *P* waves can travel through elastic solids and fluids, including the molten outer core of the Earth. *Also known as* **pressure wave**.

paleolatitude Latitude at which a rock or fossil formed; because of continental drift, this may be different from the rock or fossil's present latitude. A rock or fossil's paleolatitude may be determined from the inclination of magnetization acquired at its formation.

paleomagnetic pole Ancient position of the geomagnetic pole relative to rocks of a particular continent. If the continent has drifted since the rock formed or was magnetized, the paleopole may no longer coincide with the geographic pole: the difference indicates the amount of continental drift that has taken place. *Also known as* **paleopole**.

paleomagnetism Study of the history of Earth's magnetic field and related phenomena through measurement and analysis of the remanent magnetization of rocks, sediments and ancient fired materials.

paleopole *See* **paleomagnetic pole**.

paramagnetism Magnetic property whereby, when an external magnetic field is applied to a substance, a weak magnetization is induced in it parallel to the external field; this magnetization is lost when the inducing field is removed.

plate tectonics Theory that the Earth's surface comprises a number of rigid plates extending through the crust and into the uppermost mantle; the plates move relative to each other, driven by the process of mantle convection, new crustal material being formed at mid-ocean ridges and old material returning to the mantle at ocean trenches or subduction zones.

polar wander path/apparent polar wander path Trajectory on the surface of the Earth indicating the apparent motion of one of the paleomagnetic poles relative to the present position of the corresponding geographic pole. Differences between the polar wander paths derived from the rocks of different continents led to the acceptance of the theories of continental drift and plate tectonics.

polarity chron Major interval of the geomagnetic polarity timescale, comprising a period of predominantly normal polarity or one of predominantly reversed polarity; each chron corresponds to a major feature of seafloor magnetic anomaly profiles, and is typically around one million years in duration.

polarity reversal *See* **geomagnetic field reversal**.

pole (of Earth) *See* **geographic poles**; **geomagnetic poles**; **magnetic poles**.

pole (of magnet) Point near the end of a magnet, or on a terrella or globe, from which magnetic field lines diverge (north pole) or to which they converge (south pole).

poloidal magnetic field Magnetic field that emerges through the core–mantle boundary, and constitutes the geomagnetic field observed at and above the surface of the Earth.

portolan Medieval nautical navigation chart, showing compass bearings to be followed between sea ports.

potassium–argon dating method Radiometric dating technique important in the development of the geomagnetic polarity timescale. The radioactive isotope potassium-40 (^{40}K) makes up a small but measurable fraction of the potassium contained in many rocks; it decays radioactively to argon-40 (^{40}Ar). Comparison of the amount of ^{40}Ar trapped in a rock with the amount of ^{40}K remaining can be used to determine how long ago the rock was formed.

precession (of equinoxes) Slow periodic rotational motion of Earth's axis in space, due to gravitational forces acting between Earth and other planetary bodies.

pressure wave *See P* wave.

prime meridian Meridian from which longitude is measured; today this is the meridian through Greenwich, England. In earlier times other meridians were favored—for example, that through the Azores, where the declination was zero in about AD 1600.

radiometric/isotopic age estimation Any of several techniques used to date materials from the relative abundances of certain naturally occurring radio-active isotopes and their radiogenic decay products in the material—for example, potassium–argon.

remanent magnetization Magnetization of a material or sample over and above any induced magnetization. In other words, the magnetization retained by a sample when it is removed from any magnetic field. Remanent magnetization is carried by grains of ferri- or ferromagnetic minerals.

***S* wave** Transverse seismic wave that travels more slowly than a *P* wave, and so is felt later. *S* waves cannot propagate through fluids, and so do not enter Earth's liquid outer core. *Also known as* **secondary or shear wave**.

secondary or shear wave *See S* wave.

secular variation Gradual changes in the declination, inclination and/or intensity of the (internal) geomagnetic field that take place over time-scales ranging from tens to thousands of years.

self-reversal Rare magnetic property whereby a material acquires a stable remanent magnetization in the opposite direction from the prevailing magnetic field.

self-sustaining dynamo Dynamo in which the induced current or flow of a conductive medium generates a magnetic field that reinforces the origi-nal field inducing the current or flow.

solar wind Stream of electrically charged particles continually emitted by the sun, and flowing past the Earth at supersonic speeds.

spherical harmonic analysis Mathematical technique by which a series of measurements made on the surface of a sphere is represented in terms of a sum of wavelike functions; used in geomagnetism and other branches of physics and geophysics, such as gravity.

subduction zone *See* **ocean trench**.

sunspot Cooler region of the sun's surface appearing as a dark spot, and associated with intense magnetic fields.

terrella Solid sphere of lodestone that naturally takes on a uniform magnetization. Shown by Petrus Peregrinus and William Gilbert to have two poles, and later shown to have a dipolar magnetic field.

thermoremanent magnetization Magnetization acquired by a material as it cools through the Curie temperature of its constituent ferri- or ferromagnetic minerals.

toroidal magnetic field Magnetic field that is confined within the Earth's core and does not emerge through the core–mantle boundary; cannot be observed at the Earth's surface.

torsion balance Device employed for measuring weak forces. It consists of a bar suspended horizontally from its center of mass by a fine thread or wire. When a force is applied perpendicular to the end of the bar, the bar rotates; the angle of rotation is proportional to the force.

transform fault Boundary between two tectonic plates where the plates slide past each other. Transform faults occur mainly at ocean ridges, where they connect sections of the ridge system.

variation Original term for declination; used only rarely today.

westward drift General, predominantly westward, drift of features of Earth's magnetic field due to secular variation.

Select Bibliography

My journey through geomagnetism has led me to many hundreds of books, review and research papers, historical publications, encyclopedias and, of course, websites. In a book of this nature it is impractical to give a full bibliography: I have restricted myself to major sources of information, and books that will be accessible to the general reader who is interested in reading more widely on the subject.

De Magnete, William Gilbert, 1600; translated by P. Fleury Mottelay: Courier Dover Publications, New York, 1958

The Earth: Its Origin, History and Physical Constitution, Harold Jeffreys: Cambridge University Press, Cambridge, 1976

"Earth's Core and the Geodynamo," Bruce A. Buffett: *Science 288*, 2000

Encyclopedia Britannica: www.britannica.com

Encyclopedia of Geomagnetism and Paleomagnetism, David Gubbins and Emilio Herrero-Bervera, editors: Springer, Dordrecht, 2007

Faraday, Maxwell and Kelvin, D.K.C. MacDonald: Doubleday, London, 1964

The Fellowship: The Story of a Revolution, John Gribben: Allen Lane, London, 2005

Foundations of Modern Physical Science, Gerald Holton and Duane H.D. Roller: Addison-Wesley, Reading, Massachusetts, 1958

Fundamentals of Geophysics, William Lowrie: Cambridge University Press, Cambridge 1997

Geomagnetism, S. Chapman and J. Bartels: Clarendon Press, Oxford, 1940; second edition 1962

Latitude and the Magnetic Earth: The True Story of Queen Elizabeth's Most Distinguished Man of Science, Stephen Pumfrey: Icon Books, Cambridge, 2002

Longitude: The True Story of a Lone Genius Who Solved the Greatest Scientific Problem of His Time, Dava Sobel: Fourth Estate, London, 1995

The Magnetic Field of the Earth: Paleomagnetism, The Core, and The Deep Mantle, Ronald T. Merrill, Michael W. McElhinny, and Phillip L. McFadden: Academic Press, San Diego, 1998

"A Millenium of Geomagnetism," David P. Stern: *Reviews of Geophysics 40*, 2002

Michael Faraday: A Biography, L. Pearce Williams: Chapman and Hall, London, 1965

The Origins and Growth of Physical Science, Volume 2, D.L. Hurd and J.J. Kipling: Pelican, London, 1970

Plate Tectonics and Geomagnetic Reversals: Selected Readings and Introductions, A. Cox, editor: W.H. Freeman and Co. Ltd., San Francisco, 1973

Remarkable Physicists: From Galileo to Yukawa, Ioan James: Cambridge University Press, Cambridge, 2004

The Road to Jaramillo: Critical Years of the Revolution in Earth Science, William Glen: Stanford University Press, Stanford, 1982

The Royal Institution of Great Britain: www.rigb.org

The Royal Society: www.royalsociety.org

St. Andrews University, Scotland: www-groups.dcs.st-and.ac.uk/~history/ Biographies

Science and Civilization in China, Volume 4: Physics and Physical Technology, Joseph Needham: Cambridge University Press, Cambridge, 1962

A Source Book in Physics, William Francis Magie: Harvard University Press, Cambridge, 1935

"A three-dimensional self-consistent computer simulation of a geomagnetic field reversal," G.A. Glatzmaier and P.H. Roberts: *Nature 377*, 1995

Treatise on Geophysics, Volume V: Geomagnetism, Masaru Kono, editor: Elsevier, Amsterdam, 2007

Illustration Credits

Illustrations are listed by page number.

58   Reproduced from *The Three Voyages of Edmond Halley in the Paramore, 1698–1701* edited by Norman J.W. Thrower: Hakluyt Society, London, 1981

64   From *De Magnete* by William Gilbert

67   From *Neu-entdeckte Phaenomena von bewunderswürdigen Würkungen der Natur* by A. Doppelmayr, Nuremberg, 1774, reproduced in *Electricity in the 17th and 18th Centuries: A Study of Early Modern Physics* by J.L. Heilbron: University of California Press, Berkeley, 1979

72   Reproduced from *Magnetism and Electricity for Students* by H.E. Hadley: MacMillan & Company, London, 1924

83   Reproduced from F*oundations of Modern Physical Science* by G. Holton and D.H.D. Roller: Addison-Wesley, Reading, Massachusetts, 1958

86   Reproduced from *Foundations of Modern Physical Science* by G. Holton and D.H.D. Roller: Addison-Wesley, Reading, Massachusetts, 1958

90   (top) The Royal Institution, London

90   (bottom) The Royal Institution, London/The Bridgeman Art Library

92   Natural Philosophy Collection, University of Aberdeen

96   Science Museum/Science and Society Photo Library (ref. 10302107)

103   Courtesy of Dr. F. E. M. Lilley

109   Reproduced from *Reports of the British Association for the Advancement of Science*, vol. 5, 1836

111   (top) From *Magnetismus der Erde* by Christopher Hansteen, 1819, reproduced in *Reports of the British Association for the Advancement of Science*, vol. 5, 1836

111   (bottom) From *Magnetismus der Erde* by Christopher Hansteen

116   From a lithograph by Siegfried Bendixen, 1828

121   Compiled from a program written by Dr P.L. McFadden, Australian Geological Survey Organization, 2000

125   Reproduced from *Magnetism and Electricity for Students* by H.E. Hadley: MacMillan & Company, London, 1924

126 Reproduced from *Magnetism and Electricity for Students* by H.E. Hadley

134 Drawing by W. Alexander, reproduced courtesy of Hulton Archive/ Getty Images

145 Illustration courtesy of the author

146 Illustration courtesy of the author

147 Illustration courtesy of the author; after Inge Lehmann, 1936

149 (top) Illustration courtesy of the author

149 (bottom) Cut-away of Earth showing interior structure Mehau Kulyk/ Science Photo Library

155 Reproduced from *Comptes Rendus de L'Académie des Sciences,* series II (Earth and Planetary Sciences), vol. 328, 1999, p. 143

163 AIP Emilio Segre Visual Archives, Weber Collection

176 Courtesy of Edward Irving

179 From *Time* magazine, September 27, 1954

188 Illustration by Matthias Meyer

191 Reproduced from *Plate Tectonics and Geomagnetic Reversals* edited by Allan Cox: W.H. Freeman & Co. Ltd, San Francisco, 1973; based on an original in Raff & Mason, 1961, *Geological Society of America Bulletin 72*, pp. 1267–1270

193 Illustration by Matthias Meyer

201 Illustration courtesy of the author

205 Reproduced from *Fundamentals of Geophysics* by William Lowrie, Cambridge University Press, Cambridge, 1997, based on an original in Heirtzler et al, *Deep Sea Research 13*, pp. 427–443, 1966

206 Illustration courtesy of the author

218 Courtesy The Royal Society

226 Illustration by Matthias Meyer

228 Reproduced with permission of the Computer Laboratory, University of Cambridge, England

Acknowledgments

Researching and writing *North Pole, South Pole* has been one of the biggest challenges of my career. Among all the tasks of a university academic it is not easy to dedicate the time necessary to produce a book that is neither a text nor a research manual, but the history of a scientific quest that has spanned several millennia. It has, nonetheless, been one of the richest experiences of my life, and this is in no small part thanks to the tremendous help and support of many friends, colleagues and family members, and the wonderful people at Awa Press.

My preliminary research was carried out during study leave at the National Oceanography Center and University of Southampton, where I was hosted by Professor Andrew Roberts. I have fond memories of evenings sitting outside Andrew's office, gazing across Southampton Water towards the New Forest and discussing geophysical complexities such as precession of the equinoxes and the Chandler Wobble, not to mention our discoveries regarding the personal lives of great scientists such as Coulomb, Gauss and Sabine. Thanks also to Professors Kathy Whaler and Ken Creer of Edinburgh University for entertaining me at short notice and generously sharing their wisdom and stories on, respectively, the geodynamo and the seminal discoveries of the Cambridge paleomagnetism group in the 1950s.

I am grateful to Henry Frankel and Ted Irving for their correspondence on Jan Hospers' work, and for sharing their memories of him around the time of his death in late 2006; to Ted Evans for lively discussions of Halley's life and work; Carlo Laj for insights into the life of his "great uncle" Bernard Brunhes; Ted Lilley for enthusing me with his passion for d'Entrecasteaux's intensity measurement; and many others from whom I've gained personal anecdotes of the history of geomagnetism.

I am indebted to colleagues Darren Alexander, Paul Callaghan, Glenda Lewis, Simon Lamb, Ted Lilley, Ken Creer and Kathy Whaler who read early drafts and whose suggestions enabled me to make huge improvements, and to my husband Malcolm Ingham, who has read and advised me on every word of every draft.

Finally, were it not for publisher Mary Varnham I would never have begun this venture. She has reassured me every step of the way, and deserves a medal for guiding *North Pole, South Pole* through to publication. Thank you Mary, and Awa Press.

Index

Page references in **bold** refer to illustrations.

About the Author

GILLIAN TURNER, PHD, is a senior lecturer in physics and geophysics at Victoria University in Wellington, New Zealand. The winner of numerous awards for excellence in teaching and science communication, Turner has published over fifty articles in scientific journals. She lives in Wellington with her husband, physicist Malcolm Ingham. This is her first book.